ROUTLEDGE LIBRARY EDITIONS:
ENERGY ECONOMICS

Volume 3

ALASKAN OIL

ALASKAN OIL
Alternative Routes and Markets

CHARLES J. CICCHETTI

Routledge
Taylor & Francis Group
LONDON AND NEW YORK

First published in 1972 by Resources for the Future, Inc.

This edition first published in 2018
by Routledge
2 Park Square, Milton Park, Abingdon, Oxon OX14 4RN

and by Routledge
605 Third Avenue, New York, NY 10017

Routledge is an imprint of the Taylor & Francis Group, an informa business

British Library Cataloguing in Publication Data
A catalogue record for this book is available from the British Library

ISBN: 978-1-138-10476-1 (Set)
ISBN: 978-1-315-14526-6 (Set) (ebk)
ISBN: 978-1-138-30119-1 (Volume 3) (hbk)
ISBN: 978-0-203-73227-4 (Volume 3) (ebk)

Publisher's Note
The publisher has gone to great lengths to ensure the quality of this reprint but points out that some imperfections in the original copies may be apparent.

Disclaimer
The publisher has made every effort to trace copyright holders and would welcome correspondence from those they have been unable to trace.

ALASKAN OIL: ALTERNATIVE ROUTES AND MARKETS

by CHARLES J. CICCHETTI

Published by Resources for the Future, Inc.
Distributed by
The Johns Hopkins University Press
Baltimore and London

RESOURCES FOR THE FUTURE, INC.
1755 Massachusetts Avenue, N.W., Washington, D.C. 20036

Resources for the Future is a nonprofit corporation for research and education in the development, conservation, and use of natural resources and the improvement of the quality of the environment. It was established in 1952 with the cooperation of the Ford Foundation. Part of the work of Resources for the Future is carried out by its resident staff; part is supported by grants to universities and other nonprofit organizations. Unless otherwise stated, interpretations and conclusions in RFF publications are those of the authors; the organization takes responsibility for the selection of significant subjects for study, the competence of the researchers, and their freedom of inquiry.

This book is one of RFF's natural environments studies, which are directed by John V. Krutilla. Charles Cicchetti, a research associate at RFF at the time this study was made, is Visiting Associate Professor of Economics and Environmental Studies at the University of Wisconsin. The book was edited by Tadd Fisher. Charts and maps were drawn by Clare and Frank Ford.

RFF editors: Mark Reinsberg, Vera W. Dodds, Nora E. Roots, Tadd Fisher.

The Johns Hopkins University Press, Baltimore, Maryland 21218
The Johns Hopkins University Press Ltd., London

Library of Congress Catalog Card Number 72-11526
ISBN 0–8018–1481–2. (paper)

Preface

Good research invariably is attended by surprises and Charles Cicchetti's work is no exception. He began this study of North Slope Alaskan oil with an understandable predisposition to assume that the environmental hazards of raising the oil and transporting it to markets would exceed the net value of the resource itself. He found, however, that for this to be so the environmental costs would need to be extreme —substantially larger than a thoughtful and objective analysis of the costs and benefits of alternative sources of supply for the domestic economy shows them likely to be. An assessment of the future risks of a project conducted under conditions not previously experienced is subject, of course, to a wide margin of error, being as much speculative as objectively demonstrable. Even so, the magnitude of the difference between benefits and costs, exclusive of the environmental costs, supports Dr. Cicchetti's conclusion that exploiting North Slope oil is justifiable as a calculated risk.

The conclusion that the exploitation of these resources is a justifiable risk does not, however, remove the principal issue. It was the Trans-Alaska *Pipeline* (TAP) rather than the petroleum *exploitation* per se that initially sparked the controversy and the litigation that—even now, several years after the oil discovery—is making its way slowly through the courts toward a Supreme Court decision. The major environmental concern relates to the transportation of oil by means of a hot oil pipeline across permafrost and through a zone of active and intense seismic disturbances in the southern two-thirds of the route from Prudhoe Bay to the port of Valdez. Unavoidable leaks at terminal facilities also pose a problem for commercial fishermen and natives who depend on the Prince William Sound fisheries. But even this problem relates more to accidents during handling in transit than to the oil field operation itself.

The question naturally arises, then, whether thought was given to minimizing the environmental hazards in transporting the petroleum to market. It is in posing and exhaustively exploring this question that Dr. Cicchetti adds a dimension to our understanding of the complex ramifications of North Slope oil.

Investigating the question of markets, the author finds that the same oil has a different value, depending on the market in which it is sold. That the value differs between, say, the Pacific Coast markets and those in the Midwest and on the East Coast is a result of the characteristics of the demand in each market and the way in which the policy of oil import quotas operates to affect the supply. Midwest and East Coast markets for Alaskan (or any other) crudes have clearing prices substantially higher than those on the West Coast, even after adjusting for transportation differentials. This being the case, the possibility of bringing North Slope oil into such markets directly becomes interesting. Dr. Cicchetti demonstrates that, although a pipeline to the Midwest would require a longer overland route, the difference in costs between a Chicago and a Los Angeles delivery would not be as large as the difference in the value of oil in the two market areas. This leads him to conclude that in economic terms (exclusive of environmental considerations) an alternative route and transport mode to the Midwest or eastern market would be more attractive than TAP. Such a route would cross Canada with a different proportion of its total length traversing Alaskan tundra, depending on which of several routes studied by Dr. Cicchetti are examined. One, particularly—from Prudhoe Bay in the North to Fairbanks and thereafter using the right-of-way of the Alcan Highway to Edmonton—not only has an economic advantage over other proposed routes, but is also vastly better than TAP on environmental considerations as well.

This is a most important finding. If it is valid, is it not surprising that the oil producers engaged in the North Slope enterprise have not proposed this route and the eastern markets it would supply? Oil producers are generally assumed to be rational, profit-conscious entrepreneurs who do not voluntarily forgo returns they might as easily have grasped. Can this observation on profit motivation and the Cicchetti findings both be true? The author turns to these questions, but in doing so he drops the assumptions of classical welfare economics that underlie his early efficiency analysis and proceeds with sophisticated institutional and empirical analyses.

The simple welfare model, in which only resources, population, and technology are recognized as constraints, gives way to many modifications accommodating policy and legal factors—in fact, the entire institutional milieu governing petroleum producers in the United States. For example, TAP can be justified financially from the producers' point of view if they pick up Alaskan oil at Valdez for shipment to Japan and supply the East Coast market with Venezuelan crudes. Doing this avoids transporting any of the North Slope oil in domestic bottoms as required under the Jones Act for shipping between U.S. ports. Moreover, under an import-for-export proposal the sales to Japan would be used to increase the producers' imports under the quota system on the East Coast. Alaskan oil can be sold in Japan at the world price of around $2 a barrel; Venezuelan oil, imported in bottoms flying foreign flags, can be marketed in the New York area at nearly double that price. The opportunities to profitably exploit the institutional environment, it would seem, match those of exploiting the resource itself. This doubtless will come as some surprise to the progenitors and practitioners of simple welfare economics. But the gains to the domestic producers will be at the expense of domestic consumers, with the state of Alaska sharing in the losses. This truism follows from the analytics of simple welfare maximization.

Dr. Cicchetti thus addresses a number of important issues in addition to the central question of the Trans-Alaska Pipeline and the natural environment. His carefully researched study will repay handsomely the investment any reader will make to understand this complex of problems.

John V. Krutilla
Director
Natural Environments Program
Resources for the Future, Inc.

August 1972

Acknowledgments

I owe much to a great many individuals who gave fully of their time, opinions, data, and encouragement during the course of my work on this study. I am especially indebted to the staff of the George Washington University Computer Center for permitting me to have access to their facilities.

The Trans-Alaska Pipeline controversy was first introduced to me by James Moorman, previously an attorney for the Center for Law and Social Policy and now Executive Director of the Sierra Club Legal Defense Fund. His informed judgment and willingness to work were extremely valuable to me in the present undertaking. During the period of my research, I also benefited from discussions with several persons who have been professionally involved in this controversy. These include David Brew of the U.S. Geological Survey; Richard Corrigan of the *National Journal*; Michelle Noble, formerly with *Platt's Oilgram Price Service*; Dale Lucas of the Canadian Embassy, Washington, D.C.; Hugh Gallagher of BP Alaska, British Petroleum, Ltd.; David Anderson, former member of the Canadian House of Commons and now leader of the Liberal party in British Columbia; Edward L. Patton, President of Alyeska Pipeline Service Company; and Peter Gruenstein and Mary Tage of Congressman Les Aspin's staff.

In addition, several individuals, under considerable time pressure, generously agreed to comment on an earlier draft of this research. I am very grateful for their helpful comments and cooperation. They include John Dienelt, attorney for the Environmental Defense Fund; Arlon Tussing, economist at the University of Alaska; Ronald Merrett, economist for British Petroleum, Ltd.; William Vogeley and two members of his staff, Richard Nehring and Nicolai Timenes, Jr., of the Office of Economic Analysis, U.S. Department of the Interior; Peter Pearse, economist at the University of British Columbia; Walter Parker, former Director of the Federal Field Committee in Alaska, and Les Aspin, economist and member of Congress.

A very special thanks is reserved for several persons whose influence on my professional development and whose encouragement in this research I cannot begin to quantify. The example of Robert Haveman, Professor of Economics at the University of Wisconsin, inspired me to begin this exercise in which analysis is applied to a major public policy issue. A. Myrick Freeman III, Associate Professor of Economics at Bowdoin College, spent several sessions with me, criticizing my approach and serving as my alter ego at several important points in this analysis. He also provided valuable comments on an earlier draft. The guidance I received at the beginning of my career from Paul Davidson of the Rutgers University Department of Economics led me to employ the mix of computer techniques and the analytical method I bring to this problem. His early interest in the domestic crude oil industry is in part responsible for my own interest.

Two individuals shared their professional expertise with me in the preparation of this manuscript. Charles Lewis, geologist and environmental consultant, generously shared maps and diagrams, which he prepared and which I have utilized to compare environmentally the alternative pipeline routes. George Rogers, an economist at the University of Alaska, not only commented at length on an earlier draft, but also agreed to draw upon his own first-hand observations of the Alaskan economy in writing Appendix C. The contribution of both researchers extends far beyond my specific references to them in the text.

To my colleagues at Resources for the Future, I would like to express my gratitude for their valuable comments and encouragement. The gentle direction of John Krutilla, director of RFF's Natural Environments Program, fostered a disciplined approach while still permitting a free and exciting working environment. His aid throughout the past year contributed greatly to much of what is good in this study. Often to the detriment of his own research, he left his door ajar so that I could try out my conclusions on him before taking pen in hand. I am greatly in his debt. I also must especially thank Joel Darmstadter, Michael Brewer, Jack L. Knetsch, Orris Herfindahl, Anthony Fisher, Milton Searl, and Robert K. Davis. V. Kerry Smith unselfishly commented on each draft of this paper and contributed greatly to the present version. For typing these pages through several drafts, helping me keep track of my material, and for her encouragement, I owe a great deal to Rita Gromacki. For an exceptional and expeditious job of editing, I am indebted to Tadd Fisher.

While much of my analysis may find acceptance among those listed here, it is just as likely that there will be dissenters among this group, whose members sometimes find themselves on different sides of the North Slope oil controversy. Accordingly, the standard disclaimer about the author's sole responsibility must be emphasized in regard to my study.

I am very grateful for the understanding of my wife, Pat, and children, Colleen and Charles, when I was absent while preparing these pages and for the encouragement and comfort they gave me when my spirits were most in need of the support that only they could provide.

August 1972 Charles J. Cicchetti

Contents

List of Maps

APPENDIX MAP

List of Figures

List of Tables

APPENDIX TABLES

Introduction

In the latter part of the 1960s two seemingly unrelated events occurred that are now intertwined into one of the nation's most significant economic, environmental, and energy-related controversies. First, in the period 1968–70, successful exploratory efforts resulted in the discovery of a major oil field on the North Slope of Alaska. At the end of 1970 the American Petroleum Institute estimated the proved reserves in this Prudhoe Bay field to be approximately 9.6 billion barrels. When this field is fully developed, the amount of oil ultimately recoverable may prove to be twice as large. A field of such size makes this discovery one of the single most important in the history of the domestic crude oil industry.

Second, at about the same time that oil explorations were being undertaken in Alaska, the U.S. Congress enacted the National Environmental Policy Act of 1969 (NEPA).[1] One of the requirements of this act is that when agencies of the federal government undertake actions that might have an impact on the environment they must file an environmental impact statement (frequently called a 102(C) statement). In these statements the agencies must analyze and quantify the expected effect of the proposed action on the environment. Furthermore, when irreversible deleterious effects on the environment are found, the agencies must also consider alternatives to the proposed plan. The stipulation of this consideration has been interpreted as involving both an

[1] Public Law 91–190, 91st Congress.

environmental and economic comparison of alternative actions, including, but not necessarily restricted to, rejection of the proposed action.

After the discovery of the North Slope oil, the companies involved considered several plans for transporting the oil to market: (1) completely by water in large icebreaking tankers, (2) completely overland by a pipeline through Canada, and (3) by a combined land and water system that would require the construction of a pipeline through Alaska to a southern Alaska port, where the oil would be shipped to market by conventional tanker. Technical, economic, and legal problems, and the less than totally successful test run of the *S.S. Manhattan*, a U.S. icebreaking tanker, led to the rejection of the first alternative.

Finally, after considering the profitability, timing, costs, and legality of the two pipeline alternatives, a consortium of seven oil companies[2] sought a permit to construct a pipeline and a construction road between Prudhoe Bay in northern Alaska and the southern Alaska port city of Valdez. The proposed pipeline route is 789 miles long and the construction road is 373 miles long.[3] At the time the project was first proposed, 81 percent of the pipeline and 98 percent of the construction road were to have been built on land owned and managed by the federal government.

Several legal actions were filed against the U.S. Department of the Interior (USDI) to enjoin the granting of this permit to the oil companies. One legal action, based on the settlement of native land claims, required action on the part of Congress. This issue was greatly eased by the enactment of the Alaskan Native Claims Act in 1971. A group of Cordova fishermen initiated a second legal action to halt the proposed pipeline on the basis of the project's expected negative effect on their livelihood, which is derived from fishing in Prince William Sound.

In 1970 three environmental groups—The Wilderness Society, Environmental Defense Fund, and Friends of the Earth—filed a third

[2] The Alyeska Pipeline Service Company, owned by Amerada Hess Corporation, ARCO Pipe Line Company, BP Pipe Line Corporation, Humble Pipe Line Company, Mobil Pipe Line Company, Phillips Petroleum Company, and Union Oil Company of California. The acknowledged major ownership is in the control of the three parent companies that have found the most North Slope oil: Atlantic Richfield Company (ARCO), British Petroleum Company Limited, or BP—its U.S. subsidiary is BP Oil Corporation, which merged with the Standard Oil Company (Ohio), or SOHIO, on January 1, 1970—and the Standard Oil Company of New Jersey (Humble).

[3] The proposed road would be required from the Yukon River to Prudhoe Bay, since there is no all-weather road between those points now. Existing highways would be used between the Yukon and Valdez.

legal action, which is addressed in considerable detail in this monograph. The legal basis of the environmentalists' initial opposition was the National Environmental Policy Act of 1969, which they claimed required USDI to file an environmental impact statement before granting a permit for the proposed pipeline. Their principal concerns are the following:

1. The effect of the proposed pipeline and of production on the caribou herds that inhabit the North Slope.
2. The possibility of pipeline leaks or breaks caused by melting permafrost, seismic action, or avalanches.
3. The physical disruption of the nation's largest remaining wilderness and the effects of the construction and operation of the pipeline on this wilderness and the many rivers and streams crossed by the pipeline.
4. The consequences of "normal" tanker loading operations in Prince William Sound at the port city of Valdez and other West Coast ports.
5. The prospect of possible tanker accidents caused by collisions, earthquakes, or tidal waves.

This legal action brought the environmentalists into a direct encounter with the oil interests and set the stage for the national conflict over how to develop the massive North Slope oil field while still preserving as much as possible of the country's largest unspoiled wilderness.

These legal actions have been consolidated into one court case. The state of Alaska and the oil companies have intervened on the side of the Interior Department. A group of Canadian environmentalists, concerned about tanker traffic from Valdez passing close to the West Coast of Canada and through Puget Sound, have intervened on the side of American environmentalists, the native peoples of Alaska, and the Cordova fishermen.

The principal purpose of this monograph is to untangle the various financial, economic, and environmental issues in the controversy in a manner consistent with the analyses implicitly required by NEPA. In order to do so in a reasonable amount of depth, I will restrict my comparison of alternative actions to those that appear most likely at the time of this writing:

1. The nonproduction of North Slope oil.
2. Delay in the production of North Slope oil.
3. Approval of the Trans-Alaska Pipeline-Tanker system (TAP).
4. Approval of an all-land Trans-Canadian Pipeline system (TCP).

I will address each of the alternatives from the standpoint of (1) national interests, using a national economic efficiency objective and a benefit-cost analysis, (2) state of Alaska interests, using a tax receipts measure, and (3) oil interests, using a profitability measure. The economic analysis, developed for the TAP alternative in chapter 1, makes it possible to quantify how large the weights given to environmental damage must be in order to foreclose or postpone the development of North Slope oil (options 1 and 2). I use a computer simulation technique that permits variation in the economic parameters in order to test the sensitivity of the results.

In chapter 2, the environmental consequences of the proposed TAP alternative are considered in detail. A second objective of this chapter is the comparison of the environmental effects of alternative transportation corridors, passing overland through Alaska and Canada, with the proposed TAP alternative in order to determine the best route from the standpoint of various environmental factors. I compare five transportation systems on the basis of permafrost, seismic risk, marine spills, physiographic barriers, effect on wildlife migration, and meteorological conditions along the routes. Ranking the various transportation systems according to their relative environmental impact makes possible the identification of the implicit trade-offs that decision makers must consider. The alternative transportation systems differ, however, in other than environmental ways. These differences are considered in subsequent chapters so that the entire range of trade-offs related to the alternatives can be determined before an irreversible decision is made.

Since overland pipeline routes through Canada are longer than the proposed Trans-Alaska Pipeline, their capital costs might be expected to be greater. On the other hand, TCP routes will not require the construction of new port facilities or a new fleet of oceangoing tankers; therefore this component of capital costs will certainly be less. In chapter 3, I present a benefit-cost analysis of the alternative Canadian routes. Since pipeline routes through Canada might serve domestic markets other than those served by the proposed Trans-Alaska Pipeline—such as the Midwest or East Coast of the United States—an economic comparison of each alternative must include relative prices as well as relative costs. Chapter 3 is a summary of the economic effects of each transportation alternative on the nation, the state of Alaska, and the oil companies.

Since the economic issues in this case are extremely complex, the comparisons presented in chapter 3 are expanded in chapter 4 to include the following additional related economic factors that must be integrated into the analysis before a decision is reached:

1. The effects of splitting the oil from a Trans-Canadian pipeline between the Midwest and the West Coast of the United States.
2. Sharing the throughput of a Trans-Canadian pipeline between Alaskan and Canadian crude oil.
3. The possible effect of Alaskan crude oil on present and future price differences in different U.S. markets.
4. The economic impact of a natural gas pipeline that would share the same corridor as an oil pipeline.

In chapter 5 I consider the option of delay and compare the various regional needs of the United States over time. Since any supply and demand imbalance would have an economic impact, each alternative transportation system is compared with respect to the expected future supply and demand conditions in the domestic market it would service. When oversupply is found to exist with a specific alternative, as in the case of TAP, the effect of secondary markets on costs or prices is examined. Of particular importance in this chapter is the integration of the different plans to deal with oversupply on the West Coast. These plans are introduced into the economic analysis in order to determine their effect on national benefits, state taxes, and oil company profits. The specific plans considered for alleviating West Coast oversupply include:

1. New pipelines from the U.S. West Coast to markets east of the Rockies.
2. Selling Alaskan oil to Japan in return for increased imports on the East Coast of the United States.
3. A new pipeline in Central America for transporting Alaskan oil to East Coast markets.
4. Shipping oil around Cape Horn to East Coast markets.[4]

Finally, the conclusions of the analysis are presented in chapter 6. It is important for the reader to note that each chapter builds on previous ones in the sense that succeeding chapters add to the complexity and reality of the analysis. Accordingly, it would be incorrect to come to any conclusions based solely on the material presented in the early pages of this analysis.

[4] Since the Panama Canal is too narrow to allow passage of large oil tankers, this long route would be necessary unless a pipeline were to be built across Central America.

1

An Economic Analysis of the Trans-Alaska Pipeline

The controversy surrounding the Trans-Alaska Pipeline (TAP) extends beyond environmental matters since the proponents of this project—like those of similar development projects that will have admitted environmental consequences—point to its economic advantages. Their arguments imply that various economic benefits, if sufficiently large, can be used to justify the deleterious effects of the pipeline on the environment.

For any project of this kind, a comparison of benefits and costs can be used to illustrate the major economic issue of whether to develop or to preserve an environment in its undisturbed state. The first step, which has been used in other analyses of this kind,[1] is to discuss the environmental and other possible consequences of a development project in a qualitative manner. When possible, such effects should be quantified in nonmonetary units (see chapter 2). The second step is to convert the nonmonetary units into economic measures of environmental damages or costs in order to compare them with the benefits of the project. Generally, this conversion is not possible, given the present state of the art of benefit-cost analysis. Accordingly, a different line

[1] Testimony of John V. Krutilla and Charles J. Cicchetti, Federal Power Commission, *In the Matter of Pacific Northwest Power Company and Washington Public Power Supply System*, Projects nos. 2243/2273. Also see Krutilla and Cicchetti, "Evaluating Benefits of Environmental Resources with Special Application to the Hells Canyon," *Natural Resources Journal*, vol. 12, no. 1 (January 1972), pp. 1–29.

of analysis is suggested—one that utilizes the methodology of benefit-cost analysis but finesses conversion—and the analytical question becomes: How great do the environmental damages have to be in order to make the choice between preservation or development a matter of social indifference?

While this framework includes the important methodology of benefit-cost analysis, it has a serious disadvantage because the analyst's level of objectivity—i.e., biases resulting from his affiliations or convictions—is left unspecified. In the analysis below I use a systematic set of calculations to evaluate the economic benefits and costs of TAP. Since the economic issues are complex, I do not present initially what I believe the final benefit and cost comparisons should be. Instead I begin with the simplest case and successively introduce more complications to illustrate the effects of changing assumptions about the relevant present and future conditions in order to elucidate these many issues.

The Cost of the Trans-Alaska Pipeline

The first step in assessing the economic benefit of TAP is to determine the cost of constructing and operating this oil transportation system. Its cost depends on the amount of oil that is transported, since both the capital and operating costs are functions of the pipeline's throughput. Table 1 shows the projected throughput schedule that has

TABLE 1. Projected Alyeska Throughput Schedule[a]

Year	Barrels per day
1975	350,000
1976	900,000
1977	1,200,000
1978	1,400,000
1979–2000	2,000,000

SOURCE: Arlon R. Tussing, George W. Rogers, and Victor Fischer, *Alaska Pipeline Report: Alaska's Economy, Oil and Gas Industry Development, and the Economic Impact of Building and Operating the Trans-Alaska Pipeline,* ISEGR Report no. 31 (Fairbanks: University of Alaska, Institute of Social, Economic, and Government Research, August 1971), table IV–1.

NOTE: Richard Nehring, in *Future Developments of Arctic Oil and Gas: An Analysis of the Implications of the Possibilities and Alternatives,* Department of the Interior, Office of Economic Analysis (May 10, 1972), reported that a more rapid three-year buildup has been described by Alyeska. Under this schedule, the first and third years would average 0.6 MMb/d and 1.6 MMb/d respectively. This schedule falls between the two used in the remainder of this monograph and therefore will not be referred to again.

[a] Implied total over 25 years equals approximately 16.7 billion barrels.

been suggested by Alyeska,[2] the pipeline service company formed by the oil companies that plan to develop Alaska's North Slope oil. Alternatively, a throughput schedule of 2 million barrels per day (MMb/d) is sometimes assumed.[3] In my comparison of these two schedules, I refer to the latter one as the accelerated throughput schedule.

Capital investments represent opportunities forgone in several time periods. Therefore, a second determinant of the project's cost is the rate of time preference. There is some agreement among economists and government officials that this social rate of discount is currently 10 percent.[4] Since benefit and cost comparisons are generally very sensitive, however, to the value of the discount rate that is selected, I also use rates of 8 and 12 percent to test the sensitivity of the conclusions related to this important parameter.

The present value of the costs of a pipeline depends on the rate of utilization of the system, that is, the time stream or scheduling of throughput. This means that the analysis also depends on several interdependent factors: (1) the project's life, or time horizon; (2) the total amount of oil transported by the project; and (3) the amount of oil that is produced in any given year.

A comparison of benefits and costs can be made in the conventional manner if the amount of oil transported each year is equal and the value or benefits per barrel remains constant over time. Under such conditions the capital costs are "annualized"[5] and annual operating costs are added to calculate the annual cost. This is divided into the annual benefits, which are also assumed to be constant in each time period, to calculate the benefit-cost ratio.

[2] See note on table 1 regarding the Alyeska schedule.

[3] Arlon R. Tussing, George W. Rogers, and Victor Fischer, *Alaska Pipeline Report: Alaska's Economy, Oil and Gas Industry Development, and the Economic Impact of Building and Operating the Trans-Alaska Pipeline*, ISEGR Report no. 31 (Fairbanks: University of Alaska, Institute of Social, Economic, and Government Research, August 1971), p. 72.

[4] See testimony of Otto Eckstein and A. Harberger in *Economic Analysis of Public Investment Decisions: Interest Rate Policy and Discount Analysis*, Hearings before the Subcommittee on Economy in Government of the Joint Economic Committee, 90 Cong. 2 sess. (1968), pp. 50–81; J. A. Seagraves, "More on the Social Rate of Discount," *Quarterly Journal of Economics*, vol. 84, no. 3 (August 1970), pp. 430–50; Water Resources Council, *Proposed Principles and Schedules for Planning Water and Related Land Resources*, pt. 2, Federal Register, vol. 36, no. 245 (1971), pp. 24144–94; and U.S. Office of Management and Budget, *Discount Rates to Be Used in Evaluating Time Distributed Costs and Benefits*, Circular A–94 (1972).

[5] Otto Eckstein, *Water Resources Development* (Harvard University Press, 1965), p. 91.

The schedule shown in table 1, however, does not have constant annual throughputs, and the accumulated throughput suggested by the Alyeska and accelerated schedules differ. The annualized capital cost approach is unsuited for such circumstances. In an earlier benefit-cost study in which there was unequal physical use over time, Davidson, Adams, and Seneca[6] recommended an alternative approach—calculating the present equivalent of use and dividing it into the capital cost to find average capital cost per unit of use. In the present analysis capital cost will be on a per barrel basis.

Adopting this approach, I derive the present equivalent of pipeline utilization by using the following equation:

$$PV_{1971} = \sum_{t=1}^{T} DTP_t(365)/(1 + r)^t \qquad (1)$$

where

PV_{1971} = the present equivalent of the throughput,
DTP_t = the average daily throughput in year t,
r = the social rate of discount, and
T = the shorter of either the useful life of the transportation system or the expected production period.

Table 2 gives the present equivalent for each throughput schedule in 1971, the year for which capital costs utilized in this analysis were estimated, and 1975, the year in which the pipeline would have begun to operate if construction had started in 1971. The present equivalent barrel calculation is made with discount rates of 8 percent, 10 percent, and 12 percent. It is important to note that the present equivalent of oil is particularly sensitive to both the discount rate and the level of throughput selected. Note also that variations in both parameters have at least a proportionate effect on the present equivalent of oil calculation shown in table 2.

There have been many estimates of TAP construction costs, but at the present time neither industry nor government sources agree on any single estimate. Therefore, the strategy adopted here is to use various capital costs from \$1.75 billion to \$2.5 billion in current 1971 U.S. dollars in a simulation model that will show the sensitivity of the re-

[6] Paul Davidson, F. Gerald Adams, and Joseph J. Seneca, "The Social Value of Water Recreation Facilities Resulting from an Improvement in Water Quality: The Delaware Estuary," in Allen V. Kneese and Stephen C. Smith, eds., *Water Research* (Johns Hopkins Press for Resources for the Future, 1966), pp. 190–91.

TABLE 2. Present Equivalent of Oil Throughput

(billions of barrels)

	Alyeska throughput	Accelerated throughput
Total undiscounted throughput	16.74	18.25
Present equivalent in 1975 at 8%	7.02	8.42
Present equivalent in 1971 at 8%	5.16	6.19
Present equivalent in 1975 at 10%	5.92	7.29
Present equivalent in 1971 at 10%	4.04	4.98
Present equivalent in 1975 at 12%	5.06	6.41
Present equivalent in 1971 at 12%	3.22	4.08

NOTE: Using Nehring's revised Alyeska schedule (see note for table 1), the present equivalent in 1975 and 1971 at 10% becomes 6.42 billion barrels and 4.38 billion barrels respectively.

sults to variations in this parameter. One basis for selecting this range of estimates is shown in table 3 and its notes.[7]

For each throughput schedule and discount rate, the average capital costs per barrel for transporting North Slope oil to Valdez can be calculated by dividing the estimated capital costs (1971 U.S. dollars) by the present equivalent of oil throughput in 1971:

$$ACC = \frac{CC(1971)}{PV_{1971}} = \frac{CC(1971)}{\sum_{i=1}^{T} DTP_t(365)/(1+r)^t} \qquad (2)$$

where ACC is the average capital cost per present equivalent of oil throughput—thus ACC has units of dollars per barrel—and CC (1971) is the estimated capital cost in 1971. Table 4 shows the value of ACC for each throughput, discount rate, and capital cost case.

Another major component of cost for the TAP transportation system is the operating cost, which has been estimated to be between 12¢ and 14¢ per barrel by Tussing, Rogers, and Fischer.[8] Parker estimated the

[7] The most recent estimate utilized by the Interior Department placed the 1973 construction costs at $2.8 billion; coverting this back to 1971 for comparability and using a 10 percent discount rate implies a 1971 capital cost estimate of $2.3 billion, which is within the range utilized in this analysis. See Richard Nehring, *Future Developments of Arctic Oil and Gas: An Analysis of the Implications of the Possibilities and Alternatives*, Department of the Interior, Office of Economic Analysis (May 10, 1972).

[8] Tussing, Rogers, and Fischer, *Alaska Pipeline Report.*

TABLE 3. Breakout of Costs for the Trans-Alaska Pipeline, by Components

Component	Cost
	(*$ million*)
Haul road to be paid by oil companies	120
Approximate interest cost on $300 million pipe purchase from Japan and assumed other purchases of $100 million for 2 years at 15%	130
Initial terminal facility cost	280
Subtotal	530
Additional pumping requirements and terminal cost expansion to a storage capacity of 20 million barrels and ballast water treatment facilities in Valdez	250 to 500
Pipeline construction and acquisition cost for 800 miles at $1.2 million per mile	960
	(*$ billion*)
Total (approx.)	
Low	1.75
Medium	2.00

NOTES AND SOURCES: These figures are based on two sources:

(1) Richard A. Rice, *The North American Arctic Coast Energy Transport and Resource Distribution Problems* (Carnegie-Mellon University, Transportation Research Institute, February 1971). Rice's estimate for 1971 was $1.8 billion to $2.2 billion (p. 45).

(2) Walter B. Parker, *A Comparison of Prudhoe Oil Costs via Valdez or via a Mid-Canada Pipeline,* Federal Field Committee for Development Planning in Alaska (Anchorage, March 1971). Parker estimated the pipeline construction cost to be $1.3 billion, with added terminal construction costs of $400 million to $500 million as well as interest costs during construction. Therefore, his estimate would fall in the low range given in the table.

In *Alaskan Pipeline Report,* p. 103 (see source for table 1), Tussing, Rogers, and Fischer did not break out the costs by pipeline, pumping, and terminal categories but used a total system cost of $1.75 billion in 1971 prices, which rises to $2.17 billion when interest during construction at 10% per year is added over an anticipated four-year construction period.

Finally, in State of Alaska, *Comments on the Proposed Trans-Alaska Pipeline,* State of Alaska Report (Fairbanks: July 30, 1971), it was suggested that the range was between $1 billion and $2.5 billion, with the most likely estimate being $2.3 billion. This is consistent with Richard Nehring's estimate (see footnote 7 in the text).

pipeline component of the operating costs at 6¢ per barrel.[9] I use 13¢ per barrel as an estimate of the average operating costs. Adding ACC and the average operating costs per barrel gives the total average cost

[9] Walter B. Parker, *A Comparison of Prudhoe Oil Costs via Valdez or via a Mid-Canada Pipeline,* Federal Field Committee for Development Planning in Alaska (Anchorage, March 1971).

TABLE 4. Capital Cost per Barrel for the Trans-Alaska Pipeline and Terminal Facility

(dollars per barrel)

Discount rate	Capital cost ($ billion, 1971) 1.75	2.00	2.25	2.50
	Alyeska throughput			
8%	0.339	0.388	0.436	0.485
10%	0.433	0.494	0.557	0.619
12%	0.544	0.622	0.699	0.777
	Accelerated throughput			
8%	0.283	0.323	0.364	0.404
10%	0.352	0.402	0.452	0.502
12%	0.429	0.491	0.552	0.613

of a barrel of oil transported via TAP. Estimates of total average costs are shown in table 5 for each throughput, capital cost, and discount rate case. Table 6 is a summary of the estimates made by others of the average transportation cost for TAP.

A final component of the transportation cost is the cost of transporting the oil from the terminal city of Valdez, Alaska, to its final market destination. Table 7 shows approximate oil tanker rates for various transportation routes. I use these costs to analyze various transportation and marketing possibilities. Both the U.S. flagship and

TABLE 5. Total Cost per Barrel from the North Slope to Valdez, Assuming Constant per Barrel Operating Costs over the Life of the Project

(dollars per barrel)

Discount rate	Capital cost ($ billion, 1971) 1.75	2.00	2.25	2.50
	Alyeska throughput			
8%	0.469	0.518	0.566	0.615
10%	0.563	0.625	0.687	0.749
12%	0.674	0.752	0.829	0.907
	Accelerated throughput			
8%	0.413	0.453	0.494	0.534
10%	0.482	0.532	0.582	0.632
12%	0.559	0.621	0.682	0.743

TABLE 6. Comparative per Barrel Cost Estimates

(*cents*)

Source of estimate	Cost per barrel
Arlon R. Tussing, George W. Rogers, and Victor Fischer, *Alaska Pipeline Report: Alaska's Economy, Oil and Gas Industry Development, and the Economic Impact of Building and Operating the Trans-Alaska Pipeline,* ISEGR Report no. 31 (Fairbanks: University of Alaska Institute of Social, Economic, and Government Research, August 1971), using an 8% discount rate.	40–50
Estimates found in R. B. Norgaard, "Resource Use Conflicts Involving Oil and Gas in Alaska" (preliminary draft, Department of Agricultural Economics, University of California, Berkeley, 1971, mimeo.).	45–60
Humble Oil & Refining Company (*Oil and Gas Journal,* September 29, 1969, pp. 47–48), less 21¢ per barrel for tanker fees to Seattle. See also Richard A. Rice, *The North American Arctic Coast Energy Transport and Resource Distribution Problems* (Carnegie-Mellon University, Transportation Research Institute, February 1971), p. 19.	49–59
Cabinet Task Force on Oil Import Control, *The Oil Import Question* (February 1970). See also Charles J. Cicchetti and John V. Krutilla, "Prepared Statement on the Trans-Alaska Pipeline Environmental Impact Report," filed with the Department of the Interior, March 5, 1970. Along with its *Final Environmental Impact Statement,* the Department of the Interior prepared three economic and security volumes. In this analysis, the department used the Cabinet Task Force estimate as its "best" estimate. See Nicolai Timenes, Jr., "Analysis of Transportation Alternatives," app. C, *An Analysis of the Economic and Security Aspects of the Trans-Alaska Pipeline,* vol. 1, *Summary,* Department of the Interior (December 1971).	60

foreign flagship rates are given in table 7 because the Jones Act[10] requires that all petroleum transported between two cities in the United States be carried in U.S. flagships. This restriction is enforced irrespective of any possible resource savings that might result from utilizing less expensive foreign vessels.

Table 8 shows the total transportation costs of the proposed Trans-Alaska Pipeline and tanker operations to Los Angeles for each throughput, discount rate, and capital cost case. These estimates are based on tanker costs of 35¢ per barrel and pipeline and terminal operating costs of 13¢ per barrel.

[10] See 46 United States Code 861 et seq. (1964). Also see Kenneth W. Dam, "Implementation of Import Quotas: The Case of Oil," *Journal of Law and Economics,* January 1971, pp. 1–60.

TABLE 7. Approximate Tanker Tariffs

(dollars per barrel)

	Foreign flag		
	Future[a]	Present	U.S. flag
Persian Gulf to Los Angeles	0.44	0.74	–
Valdez to Japan	0.20	0.34	–
Valdez to Los Angeles	0.11	0.19	0.35
Valdez to Panama to New York (pipeline across Panama)	0.32[b]	0.53[b]	0.65 to 1.06[c]
Valdez to New York around Cape Horn	0.66	1.11	1.32 to 2.22[c]
Seattle to Los Angeles	–	–	0.15

SOURCES AND NOTES: Based on rates utilized by the Cabinet Task Force on Oil Import Control in *The Oil Import Question* (February 1970) and a simplified linear approximation. These estimates are also consistent with the values used and discussed by Arlon R. Tussing, George W. Rogers, and Victor Fischer, *Alaska Pipeline Report: Alaska's Economy, Oil and Gas Industry Development, and the Economic Impact of Building and Operating the Trans-Alaska Pipeline*, ISEGR Report no. 31 (Fairbanks: University of Alaska, Institute of Social, Economic, and Government Research, August 1971), p. 106, fn. 14. It should be pointed out that the range of tanker costs for the important Valdez-to-Los Angeles run in the Tussing report range from 20¢ to 46¢ and bracket the 35¢ value shown above, with the most probable estimate around 30¢. In *An Analysis of the Economic and Security Aspects of the Trans-Alaska Pipeline* (December 1971), the Department of the Interior utilizes 30¢ for an average Valdez-to-West Coast rate. However, Nehring updates the estimate of tanker capital costs from $1 billion to $1.7 billion, including $0.2 billion for new West Coast port facilities This would increase the costs per barrel of the proposed North Slope-to-Los Angeles transportation system by 14¢ to 17¢ per barrel, using the original Alyeska and accelerated throughput schedules respectively, and an assumed tanker life equal to the pipeline life. See Richard Nehring, *Future Developments of Arctic Oil and Gas: An Analysis of the Implications of the Possibilities and Alternatives*, Department of the Interior, Office of Economic Analysis (May 10, 1972).

[a] Estimate based on improved economies due to the use of supertankers.

[b] Terminal and pipeline costs must be added. Assume these equal 25¢ per barrel; therefore, estimates of 57¢ and 78¢ can be determined by adding this pipeline cost to the tanker costs of 32¢–53¢ per barrel.

[c] Assume U.S. flagships would cost twice as much per barrel for operating and construction costs.

A Benefit-Cost Analysis

Production and transportation costs are not the only variables that must be considered in order to evaluate the economic merits of TAP. By analyzing the price at which the oil is sold, as well as the system's costs, several relevant decision-making parameters can be determined:

$$\text{Tidewater price per barrel} = \text{market price per barrel} - \text{tanker costs per barrel} \quad (3a)$$

$$\text{Wellhead price per barrel} = \text{tidewater price per barrel} - \text{pipeline costs per barrel} \quad (3b)$$

$$\text{Wellhead value or net revenue per barrel} = \text{wellhead price per barrel} - \text{production and gathering costs} \quad (3c)$$

Equations 3b and 3c are especially important for determining the economic desirability of the proposed system. First, the state of Alaska will collect revenues of approximately 20 percent of the wellhead price. Second, the net revenue per barrel represents the return on the oil companies' initial investment of more than $1 billion in exploration and lease purchases before corporate income taxes are imposed.[11]

TABLE 8. Total Cost per Barrel to Los Angeles from the North Slope via the Trans-Alaska Pipeline and U.S. Flagship

(dollars per barrel)

Discount rate	Capital cost ($ billion, 1971) 1.75	2.00	2.25	2.50
	Alyeska throughput			
8%	0.819	0.868	0.916	0.965
10%	0.913	0.975	1.037	1.099
12%	1.024	1.102	1.179	1.257
	Accelerated throughput			
8%	0.763	0.803	0.844	0.884
10%	0.832	0.882	0.932	0.982
12%	0.909	0.971	1.032	1.093

NOTE: The estimates in this table are based on tanker costs of 35¢ per barrel and pipeline and terminal operating costs of 13¢ per barrel.

Returns to the state of Alaska and the oil companies are important economic considerations, but they are excluded from the benefit-cost

[11] In the analysis that follows, the before-corporate-income-tax calculations will be utilized, since the special tax provisions of the crude oil industry make it nearly impossible to estimate the after-tax return. Furthermore, as recently as 1968, when the oil depletion analysis was set at 27.5 percent rather than the present 22 percent, the major oil-producing companies in the United States paid an *average* corporate income tax of only 7.1 percent. Two of the companies that have major Alaskan holdings, Standard of New Jersey (Humble) and Atlantic Richfield, paid 9.7 and 1.2 percent respectively in 1968. See *Congressional Record,* daily ed., August 5, 1969, p. E–6654.

calculus. Both returns represent monetary transfers between different individuals in the same society. They are important determinants of the preferred choice of each group. However, in calculating national benefits, which economists call economic efficiency, it is not the distribution of the TAP benefits that is important, but the size of the total resource savings and other benefits.[12] One method of calculating the economic efficiency benefits of producing and transporting oil on the North Slope is to determine the costs of an alternative means of supplying oil to the same market. Thus net benefits can be defined as follows:[13]

$$b = (B^i_{NS} - C^i_{NS}) - (B^i_A - C^i_A) - \bar{X} \tag{4}$$

where

b = net benefits of the proposed production and transportation of oil from the North Slope,

B^i_{NS} = gross benefits of oil produced on the North Slope when it is utilized by consumers in market i,

C^i_{NS} = real cost of producing and transporting North Slope oil to consumers in market i,

B^i_A = gross benefits of an alternative source of oil in market i,

C^i_A = real cost of producing and transporting an alternative source of oil to market i, and

$\bar{X}$ = a row vector of other considerations, including environmental costs, whether quantifiable or not in dollar terms, as well as other real economic benefits and costs, such as national security considerations, etc.

If both alternatives serve the same market (i) and the quality of substitute crude oils are similar, then consumers would be indifferent to the two alternative sources of supply. That is, the gross benefits of either source are identical, $B^i_{NS} = B^i_a$. Furthermore, if it is assumed—as the Interior Department has assumed implicitly in its economic analysis—(1) that there are no economically superior (i.e., lower-cost)

[12] An exception would occur if, for example, society valued a dollar received by an oil company stockholder or the state treasury differently from a dollar kept by an oil consumer. Under such circumstances *real* economic benefits and costs would be generated as external effects of the project. Since sufficient information is lacking for the evaluation of such distributional effects, they will not as a rule be considered in this analysis.

[13] For a more detailed discussion of this approach, see Peter O. Steiner, "The Role of Alternative Cost in Project Design and Selection," *Quarterly Journal of Economics* 79(1965): 421–22.

North Slope (NS) to i transportation alternatives, (2) that market i is the highest-valued final destination for North Slope crude, and (3) that $\bar{X}i$ is zero, then the net benefits of the North Slope oil would reduce to

$$b' = C_A - C_{NS}. \tag{5}$$

Equivalently, the ratio of the costs of the best alternative to the costs of the proposed development project can be used to determine the conventional benefit-cost ratio of the development project when the three assumptions hold.

$$\mathrm{B/C} = \frac{\text{Cost of alternative oil}}{\text{Cost of North Slope oil}}. \tag{6}$$

A further test must be performed before development is approved, namely, the gross benefit of the oil (or value in consumption) must also exceed the costs, subject to

$$B_{NS} = B_A \geq C_{NS}. \tag{7}$$

A short digression to point out some salient features of the U.S. crude oil industry is necessary in order to determine the costs of alternative sources of supply for the Los Angeles or West Coast market area. For a discussion of these features, see appendix A—an analysis showing that the Persian Gulf would be the most probable alternative foreign source of supply to U.S. West Coast markets if TAP were not built. Further, it is not likely that domestic oil from developed West Coast sources of supply will meet all the growing demand on the coast in the future.[14] This conclusion is subject to the following assumptions: (1) the Trans-Mountain Pipeline from Edmonton to Puget Sound is not expanded, (2) petroleum District V (West Coast) prices remain at their current levels, and (3) there is no appreciable change in interdistrict flows of oil in the United States.

Appendix A includes a framework for developing a computational model to measure the net benefits of TAP. I make the following additional assumptions to keep the present analysis similar to that im-

[14] In Nehring, *Future Developments of Arctic Oil*, the possibility of increased imports from Ecuador, Peru, and Indonesia, as well as from domestic sources, is discussed. If locational savings from these foreign sources are used to increase the taxes collected by the countries, such alternatives would not affect the benefit-cost calculations that I base on a Persian Gulf alternative. If some real savings are possible, however, then I have systematically biased my calculations in favor of the proposed pipeline development.

plicitly undertaken by the Department of the Interior, but I will relax the first two later in the analysis:

1. The quantity of oil demanded at the present price in year *t* will equal the present supply plus the proposed throughput of the Trans-Alaska Pipeline in year *t*.
2. As now, foreign oil, rather than higher-cost domestic oil, would be the new source of supply for the West Coast if TAP were not built.
3. The average cost of producing and transporting a barrel of oil is equal to the marginal cost of a barrel of oil in each producing area; that is, supply functions for crude oil are perfectly elastic in certain ranges.

The Cost of the Alternative Source of Supply

The next step in the economic evaluation of the proposed pipeline is to determine the actual costs of foreign crude; this will serve as a benefit estimate and be compared with the proposed pipeline costs presented earlier. To calculate the real cost of a foreign alternative to TAP on the West Coast, a specific reference crude, Iranian Light 34° API,[15] has been selected. This cost is derived in table 9. One important difference in determining the real cost of this alternative source of oil to the United States must be reemphasized. Taxes paid to foreign governments represent real costs for the United States, while taxes paid to state governments within the country are monetary transfers. The reason is that payments made by oil companies to the state of Alaska reduce the welfare of oil company owners but benefit the taxpayers in Alaska. As indicated earlier, such transactions are usually considered transfers of income within an economy rather than real resource costs. This is not the case[16] when tax payments are made to residents of another country. At some point in the future such tax receipts may be used to demand real goods or services from the U.S. economy and would thus represent real costs to the country. Therefore, unless a world efficiency or benefit-cost point of view, which took into account comparative advantages as opposed to a national efficiency

[15] Degrees API (American Petroleum Institute) is the standard representation of the specific gravity of the particular grade of crude oil.

[16] This conclusion is not dependent on the fact that tax credits can be applied against federal corporate income taxes when royalties are paid to oil-producing countries. This is an equity consideration between taxpayers and oil company stockholders.

TABLE 9. Average Costs for Persian Gulf Oil Delivered to U.S. Coastal Ports

(dollars per barrel)

	Iranian light 34° API (1.4% sulfur)	
	1971	1975
Production[a]	0.10	0.11
Payments to foreign governments[b]	1.11	1.27
Other costs[c]	0.30	0.30
FOB arm's length price	$1.51	$1.68
Transportation costs[d]	0.74	0.45
Total cost to U.S. (delivered)	$2.25	$2.13
U.S. tariff	0.105	0.105
Delivered price	$2.36	$2.24
Average	$2.30	

[a] See M. A. Adelman, *The World Petroleum Market* (Johns Hopkins University Press for Resources for the Future, forthcoming). Adelman's calculation is best viewed as an overall average for the Persian Gulf and not necessarily for the Iranian Light 34° API that is used in this analysis.

[b] See *Platt's Oilgram Price Service*, August 25, 1971; and *Oil and Gas Journal*, June 21, 1971, for crude price postings in 1971 and 1975. Payments to foreign governments are based on a calculation of 49% of the posted price that has been approximately determined by Mikesell for Iran. See Raymond F. Mikesell, *Foreign Investment in the Petroleum and Mineral Industries: Case Studies of Investor-Host Country Relations* (Johns Hopkins Press for Resources for the Future, 1971), especially p. 247.

[c] These costs include all other transportation and gathering costs and payments made to determine the FOB arm's length price. This FOB price was estimated to be $1.30 in New England Council and others, "Rebuttal Statement of the New England Parties on Oil Import Controls to the Cabinet Task Force on Oil Import Control," submitted to the Cabinet Task Force, August 15, 1969. At the old posting of $1.80 and Adelman's production costs of 10¢, this would mean a difference of 30¢, which I denote as "other costs."

[d] I show a high and low cost for transportation to be consistent with the Cabinet Task Force calculations.

perspective, were to be adopted, such a levy by oil-producing countries would have to be considered real costs to the importing economy.[17]

Tariffs paid to the U.S. Treasury (approximately 10.5¢ per barrel) are monetary transfers. Accordingly, subtracting U.S. taxes and including taxes paid to producing countries gives $2.13 as the real cost of a barrel of oil in 1975—using Iranian Light (34° API, 1.4 percent sulfur)

[17] For a more detailed discussion of this issue and its importance, see A. Myrick Freeman III, "An Evaluation of the Final Environmental Impact Statement on Proposed Trans-Alaska Pipeline and an Analysis of the Economic and Security Aspects of the Trans-Alaska Pipeline" (submitted to the Department of the Interior, May 5, 1972).

as a reference crude. An additional adjustment is necessary to make the Iranian Light reference crude comparable to the quality of oil reported to have been found on the North Slope.[18] Using an adjustment factor of 1.5¢ per degree API means that the average cost of a heavier reference crude would be $2.01 per barrel.[19]

Additional Cost and Price Data

In order to calculate the state tax revenue and the net revenues of the oil companies, some additional information is necessary. Table A-5 of appendix A shows the prices of crude oil in various world markets. When available, the prices are shown for oil similar in quality to the oil that has reportedly been found on the North Slope.[20] Moreover, in some instances the sulfur content of the crude oil is given. Sulfur content is important in determining price because of current air pollution laws and because additional costs are borne by refiners when sulfur is present.

The wellhead price can be determined by subtracting from the price of oil in Los Angeles[21] ($3.17 for 26°–26.9° API) the transportation costs of TAP in each throughput, discount rate, and capital cost case. Tax (royalty plus severance) receipts in Alaska are equal to approx-

[18] See Tussing, Rogers, and Fischer, *Alaska Pipeline Report,* p. 106. In this report a crude of 26°–26.9° API is used to represent the crude that has been found in Alaska.

[19] *Platt's Oilgram Price Service, Crude Oil Supplement,* vol. 49, no. 164–B (August 25, 1971). An adjustment of 12¢ is based on the difference in specific gravity (34° API − 26° API = 8° API) and the 1.5¢ per degree API (1.5¢/degree API × 8° API = 12¢).

[20] See fn. 18.

[21] There is a discrepancy in the price used for a barrel of oil in Los Angeles in Interior Department studies and in the studies performed in Alaska. The Interior Department's West Coast price is consistently higher, since the department assumes that North Slope crude is lighter than the 26°–26.9° API utilized by the state of Alaska and since it adds higher gathering charges in California to the posted price of coastal California crude. I utilize the lower West Coast price of $3.17 used by the state rather than the department's price of $3.31. For a discussion of these and other price issues, see Charles J. Cicchetti and John V. Krutilla, "Comments on the Final Environmental Impact Statement for the Proposed Trans-Alaska Pipeline and the Complementary Analysis of the Economic and Security Aspects of the Trans-Alaska Pipeline" (submitted to the Department of the Interior, May 5, 1972); Tussing, Rogers, and Fischer, *Alaska Pipeline Report*; and Nicolai Timenes, Jr., "Analysis of Transportation Alternatives," app. C, *An Analysis of the Economic and Security Aspects of the Trans-Alaska Pipeline,* vol. 1, *Summary,* Department of the Interior (December 1971). Also see app. J., vol. 2, *Supporting Analysis,* of *Analysis of Economic and Security Aspects.*

imately 20 percent of this wellhead price. Furthermore, if the average production and gathering cost of 24.5¢ per barrel (based on the estimates shown in table 10) plus the tax paid to the state of Alaska are subtracted from the wellhead price, the net revenue before corporate income tax can be determined.

Net Benefits, Taxes, and Net Profits

Table 11 is a summary of the results of several important economic variables for the various cases outlined above. The stub shows the capital cost and throughput schedule that defines the case under consideration. Column 1 shows the wellhead price for each case when a 10 percent rate of discount is utilized.

If the effect of using discount rates of 8 percent or 12 percent rather than 10 percent is important, it may be determined by referring to table 8. In general, if the Alyeska throughput schedule is utilized an 8 percent discount rate will mean a 10¢–12¢ higher wellhead price per barrel and a 12 percent discount rate will mean a 10¢–12¢ lower wellhead price per barrel. Variations in the rate of discount have a similar effect on the accelerated schedule.

Column 2 shows the present value of taxes that the state of Alaska would receive in each case. Added to the nearly $1 billion already paid to Alaska in recent lease sales, the present value of the crude oil found on the North Slope is approximately $3 billion. Column 3 shows the present value of net profits that the oil companies would receive in each case after taxes are paid to the state of Alaska, but before federal corporate income taxes, if Los Angeles is used as the principal market area.

TABLE 10. Estimated Wellhead Costs for North Slope Oil Fields

(*cents*)

Source	Wellhead cost
M. A. Adelman[a]	18–26
Charles Norman[b]	23.5–27
Cabinet Task Force[c]	36 (including 8¢ royalty to state)
Average value	24.5

[a] M. A. Adelman, "Significance of Shifts in World Oil Supplies" (paper presented at the 20th Alaska Science Conference, University of Alaska, Fairbanks, August 1969).

[b] Charles Norman, "Economic Analysis of Prudhoe Bay Oil Field, Alaska" (master's thesis in economics, Massachusetts Institute of Technology, July 1969).

[c] See Cabinet Task Force on Oil Import Control, *The Oil Import Question* (1971), p. 226.

TABLE 11. Wellhead Prices, Present Value of Taxes, Net Profits, and Benefit-Cost Ratios at a Discount Rate of 10 Percent for the Trans-Alaska Pipeline and U.S. Flagship to Los Angeles

Capital cost and throughput schedule, Trans-Alaska Pipeline[a]	(1) Wellhead price (*U.S. $*)	(2) Present value of taxes (*bil. 1971 U.S. $*)	(3) Present value of net profits (*bil. 1971 U.S. $*)	(4) Benefit-cost ratio[b] (1975 OPEC prices)	(5) Benefit-cost ratio (conservative)
$1.75 billion					
Alyeska	2.26	1.82	6.3	1.94[c]	2.46[d]
Accelerated	2.34	2.32	8.1	2.12	2.71
$2.0 billion					
Alyeska	2.20	1.78	6.1	1.81	2.31
Accelerated	2.29	2.28	7.9	2.01	2.55
$2.25 billion					
Alyeska	2.13	1.72	5.9	1.70	2.17
Accelerated	2.24	2.22	7.7	1.89	2.41
$2.50 billion					
Alyeska	2.07	1.68	5.7	1.61	2.05
Accelerated	2.19	2.18	7.5	1.80	2.29

NOTE: With discount rates of 8 and 12 percent the figures for the Alyeska throughput schedule ($1.75 billion capital cost) are:

8%	2.35	2.42	9.7	2.16	2.75
12%	2.15	1.38	5.5	1.72	2.20

[a] Capital costs are given in 1971 U.S. dollars.

[b] The following algorithm was used to calculate all the benefit-cost ratios in this analysis: the cost of the alternative less North Slope production costs was divided by the total costs per barrel shown in table 8 for each capital cost case. A copy of the computer program used for these and the succeeding calculations presented in this analysis can be obtained from the author on request.

[c] This benefit-cost ratio was determined by dividing the cost per barrel of a foreign alternative less North Slope production costs by transportation costs as shown in Table 8, that is: $(\$2.01 - 24.5¢)/\$0.913 = \frac{\$1.765}{\$0.913} = \$1.94$.

[d] This benefit-cost ratio was calculated by dividing ($2.50 − 24.5¢) by $0.913 = $2.255/$0.913 = $2.46.

Columns 4 and 5 give the ratios of benefits to costs for each case. Column 4 is based on expected future tanker rates and on posted prices now agreed to for 1975. Under these conditions a real cost in California of $2.01 per barrel is used for the foreign alternative.

In column 5, a more conservative benefit-cost calculation is displayed. Its purpose is to determine the sensitivity of the benefit-cost calculation to increased costs for foreign oil. Two types of cost adjustment are made. First, it is assumed that taxes paid to the exporting country will increase by 20¢ per barrel. Recent successes by the Oil Petroleum Exporting Countries (OPEC) indicate that some increase in payments or other costs would be likely. Second, it is assumed that if tanker rates do not decline in the future an additional 29¢ per barrel must be added to the average cost of the foreign alternative source of supply. One reason for this adjustment is that present foreign tanker rates are based on tankers utilizing the same technology as that assumed for the U.S. flagships that would be used to transport oil between Valdez and Los Angeles. Both adjustments increase the average cost of the alternative from $2.01 per barrel to $2.50 per barrel. The higher benefit-cost ratios in column 6 reflect this adjustment.[22]

The benefit-cost ratios in table 11 are important to any consideration of whether or not TAP should be built. For a decision to be made to stop the construction of TAP, society must judge the benefits forgone from the cost saving implicit in the benefit-cost ratios in table 11 to be less than any excluded net external costs that would accompany the construction of the project. $\bar{X}$ designates such external effects in equation 4. Thus far they have been assumed to sum to zero.

For the middle range of the capital costs, table 12 shows how big these omitted net external costs must be to warrant a decision not to build TAP. Since the proposed project involves a stream of benefits, the present values given in the table for each throughput schedule are most important. These values can be compared with any net external costs included in $\bar{X}$ and as yet unquantified. Accordingly, given all the assumptions above, the omitted external effects must have a net present value greater than or equal to the present value of development benefits (which range between $3 billion and $6.6 billion) before it is possible to select a preservation or nondevelopment option as socially optimal. In the following discussion several of the asumptions that were made to reach this conclusion (and that may be unrealistic, if not wrong) will be relaxed in order to test the sensitivity and meaning of this conclusion.

Qualifying Factors: Some Spillover Effects

Some of the major adverse side effects (external costs) of the TAP system are environmental. These are discussed in chapter 2; they were not included in the benefit-cost calculations, but it is unlikely that these environmental costs would be zero.

[22] Note that Nehring, in *Future Development of Arctic Oil,* calculated a foreign cost in 1975 of $2.80. After removing 30¢ in oil company earnings (since U.S. companies would probably produce all foreign oil imported into the United States), his price is identical to the $2.50 used in my analysis and labeled conservative. In much of my analysis, I will use this conservative price in order to avoid possible biases against development. The analysis is implicitly biased against preservation, however.

TABLE 12. Net Benefits of the Proposed Trans-Alaska Pipeline, Assuming All External Effects Are Zero (Capital Cost, $2.25 Billion, 1971 Dollars; Discount Rate, 10 Percent)

	Present OPEC prices, future tanker costs[a]	More conservative: higher OPEC prices, present tanker costs
	(------- *1971 dollars* -------)	
Per barrel benefits		
Alyeska	0.73[b]	1.22[c]
Accelerated	0.84	1.33
	(--- *billions of 1971 dollars* ---)	
Undiscounted benefits, assuming 15 billion barrels are produced	11.0	18.3
Present value: Alyeska throughput schedule	2.9	4.9
Present value: accelerated throughput schedule	4.1	6.6

[a] Estimates based on improved economies due to the use of supertankers.

[b] Determined by subtracting production costs of North Slope oil (24¢) and transportation costs ($1.04) from the real cost per barrel of the foreign alternative ($2.01), i.e., $2.01 − 24¢ − $1.04 = 73¢ per barrel.

[c] In this case the more conservative $2.50 per barrel is used for the foreign cost alternative.

The challenge of producing North Slope oil, building a pipeline, and transporting hot oil[23] through one of the nation's last remaining vast wilderness areas is great. Special engineering problems will arise in transporting hot oil through permafrost and in pumping oil through a region of heavy seismic activity for approximately two-thirds of its overland route. Additionally, there probably will be environmental damage caused by transporting oil in oceangoing tankers across the hazardous waters of Prince William Sound at the southern terminus of the Trans-Alaska Pipeline. Tanker activity in Puget Sound and in the harbors of California port cities will also have an effect on the environment of these areas. Environmental effects are sometimes uncertain (e.g., seismic activity) and are usually difficult to quantify in monetary terms. It is therefore difficult under the present state of the art to include such environmental damages in the benefit-cost ratio.

Another problem is that changes in the environment may result in damages for many years; they may be irreversible. Therefore, to meas-

[23] The oil produced in Alaska will be about 150° F. when it is piped to Valdez.

ure damages only in the present would be to greatly underestimate them. Furthermore, society may be willing to pay greater amounts in the future to avoid the same environmental loss. As population grows and real income, education, and leisure time increase, the value of preserving the Alaskan environment for future generations of Americans may increase.[24] This means that the present annual value of such environmental resources could be small relative to the development benefits, and preservation could still be justified if the rate of appreciation in the annual value of the environmental resources is substantial.[25]

A similar adjustment to increase the relative preservation benefits should be made if society is expected to value crude oil less in the future than in the present. While present rates of growth in the consumption of crude oil may make such an expectation unlikely in the near future, a longer-run view of energy consumption patterns could prove otherwise. First, technological change affecting oil shale, coal, and the tarsands of Lake Athabasca may make these potential sources of petroleum a cheaper alternative source of petroleum in the future (assuming that any environmental problems associated with these alternatives can also be overcome or would represent a lesser environmental cost). Second, through increased research and development, advances in breeder reactors may result in a safe and cheap source of energy in the future. Finally, more esoteric sources of energy, such as fusion reactors and magnetohydrodynamic and geothermal sources of energy may be developed in the decades ahead.

Long-run changes in energy supply sources are highly uncertain and depend in large part on the amount of resources dedicated to research and development. Of course, a decision to permanently cap the important North Slope oil fields undoubtedly would spur research and development efforts to find other solutions to U.S. energy supply problems—a consideration that should be given appropriate weight. The negative supply effects of non-development must also be considered, however. Production on the North Slope may result in additional oil and gas discoveries. Thus the estimate of production benefits derived above would be underestimated.

National security is sometimes used as a major argument for building TAP. Proponents assert that constructing the pipeline would reduce

[24] Charles J. Cicchetti, "A Multivariate Analysis of Wilderness Users in the United States" (paper presented at the Natural Environments Workshop, Missoula, Montana, August 1971).

[25] See Anthony C. Fisher, John V. Krutilla, and Charles J. Cicchetti, "The Economics of Environmental Preservation: A Theoretical and Empirical Analysis," *American Economic Review*, September 1972.

U.S. dependence on a foreign source of supply. This beneficial side effect was omitted in the discussion of the components of the benefits because there is a possible offsetting factor that also must be considered. If development of North Slope oil today merely postpones the time of ultimate dependence on foreign oil, using up U.S. resources during a period of relative peace may cause the nation to find itself short of oil reserves in the future. This would be an inferior position from the standpoint of both national and financial security. Thus the United States might find itself subject to the arbitrary whims of oil monopoly states, as Western Europe and Japan find themselves today. If such a situation were likely in the future, postponement of the development of the North Slope oil accompanied by the importation of oil from Venezuela and the Persian Gulf might become socially optimal, since an oil-producing country with capped reserves would be in a stronger position to negotiate prices when it purchased oil. In other words, if prices of foreign oil are expected to increase faster than any increase in the costs of producing and transporting North Slope oil in the future, it may be socially optimal to postpone production to some future point in time in order to maximize the present value of North Slope production.

There are two possible exceptions to this national security consideration. First, it is not known what the effect on future exploration for oil in the United States would be if this most important discovery were to be left capped, but it could be quite significant. Actually, the United States does cap some of its oil for national security purposes. These fields are called Naval Petroleum Reserves, and the largest is located in Alaska to the west of the Prudhoe field. But if the Prudhoe field were to remain capped after the extensive private investment made thus far, it would have a detrimental effect on the domestic crude oil industry. Second, price negotiations are now conducted between oil companies and the producing countries; therefore, capped national reserves may prove quite ineffectual in bargaining sessions between individual oil companies and OPEC. Finally, requiring the state of Alaska to forgo expected tax benefits, whose present value is nearly $2 billion, in favor of residents of the other forty-nine states may be questioned on equity grounds even if such a decision were economically efficient.

Furthermore, when oil production is contemplated in a particular section of the United States, current institutional constraints in the domestic crude oil industry must be considered. At the present time oil could be imported to both the East Coast and West Coast from the

Persian Gulf at approximately the same cost. The price of oil in New York, however, is approximately 90¢ per barrel more than a similar barrel of oil sold on the West Coast (if 26.0°–26.9° API North Slope crude is used as a reference). There are two related reasons for this. First, the Mandatory Oil Import Quota Program is not uniformly applied in all regions of the country. The amount of imports that can be brought into Districts I through IV (states east of the Rockies) is 12.2 percent of domestic production in those regions. On the other hand, imports into petroleum District V (the West Coast) are determined on the basis of expected demand and supply conditions at the historical price per barrel of crude oil. Accordingly, if the gap between District V's producers and consumers is an expected 500,000 barrels per day, then imports will be allowed at the level of 500,000 barrels per day.

The effect of this program is full protection for producers in the western part of the country. Producers in the eastern half of the United States are protected in another way. The amount of oil that can be produced in any given field (and for that matter in any given well) is regulated by an interstate compact of regulatory agencies. The effect of this system of prorationing coupled with control of imports (as a percentage of production) is that supplies are curtailed from both domestic and international production for consumers in the eastern part of the United States, with a concomitantly higher price.[26] Given the present executive decisions concerning oil import quotas, any North Slope oil transportation system that favors the West Coast and excludes other systems that would tend to equalize the present regional differences should be studied very carefully. An issue of equity, as well as of economic efficiency, is raised when the question of where North Slope oil should be brought to market is considered. I will defer this issue of equity, however, until later chapters.

A discussion of equity suggests the possibility of alternative pipeline routes overland through Canada. As a qualification of the benefit-

[26] This price differential can be determined by using a New York price of $4.06 per barrel and a Los Angeles price of $3.17 per barrel at 26.0°–26.9° API. When lighter crudes are used for reference purposes this difference will decline somewhat (e.g., the difference at 30° API is approximately 60¢ per barrel). One reason for this is that heavier crude, such as 26.9° API, will not efficiently yield as much of the lighter petroleum products, such as gasoline and jet fuel, as lighter crudes (e.g., 30° API). West Coast consumers demand a larger proportion of lighter products than consumers east of the Rockies who proportionately use more of the heavier petroleum products for such things as home heating. An additional factor is that heavier crudes are often high in sulfur content, thus reducing their value since increased refining costs are necessary.

cost analysis above, such a consideration is important in several ways. For example, a Canadian pipeline may have greater net benefits, including reduced environmental conflicts. Such an outcome would not affect an analysis of whether a pipeline is a better way to supply the West Coast than to import a similar quantity of crude oil from the Middle East. If the net benefits of supplying another portion of the United States were found to be greater, however, then a decision to build TAP would not be economically efficient. If this occurred and the benefit-cost ratios were corrected for the omission of this opportunity benefit forgone, they would fall below unity.

Later I will discuss this qualification in detail. But it is important to point out that it will not affect the decision of whether or not to produce the oil on the North Slope. It relates merely to the market area that should be supplied by North Slope oil and, therefore, to the selection of the best pipeline route.

To obtain the transportation cost estimates, I introduced a final qualification—the implicit Interior Department assumption that demand on the West Coast would be able to absorb the full capacity throughput of the Trans-Alaska Pipeline. This assumption, which I will return to later, may be unrealistic. Even with a projected rate of growth of demand of 4 percent on the West Coast, and assuming no foreign imports to the West Coast and that oil produced in the other petroleum District V states would remain constant, by 1980 TAP would carry approximately 600,000 barrels of oil daily that would be in excess of the demands of District V at the current price per barrel. If this oversupply were dealt with as in other parts of the country by using market demand prorationing agreements, the effect would be to sharply increase the cost per barrel of TAP oil. The average costs used above are sensitive to the percentage utilization of the pipeline's total capacity. While the likelihood of prorationing is small, excess supply and its effect on the efficiency of development and on route selection should be considered very carefully.

Thus far, then, the nonstochastic simulations I have performed show that the present value of net benefits to the nation of producing North Slope oil would probably be in the range of $3 billion to $6.6 billion before possible environmental costs, national security benefits and costs, and possible changing values of crude oil are included. The weights attached to the latter three items are likely to be the source of major debate. But—if the assumptions implicitly made by the Interior Department and used in deriving these estimates are valid—my calculation sets the range of the value of benefits that must be matched by adverse side effects if capping the North Slope oil field is to be recommended as a socially optimal decision. Throughout the remainder of this analysis I will examine these assumptions in some detail.

I have estimated the present value of expected tax revenues to Alaska to be approximately $2 billion and profits before corporate income taxes to be approximately $6 billion to $8 billion. In the remaining chapters, when I consider various alternative pipeline routes and possible demand and supply imbalances at greater length, I will concentrate the analyses on the sensitivity of these tax and profit estimates in addition to the benefit-cost efficiency criteria.

2

Some Environmental Considerations

Much has been written about how the production and transportation of North Slope oil will necessitate invading areas previously undisturbed by man and his technology. Emphasis has been given to the dangers of oil spills and leaks and the possible peril to Arctic flora and fauna. But the environmental issues in the Alaskan oil controversy are both highly complex and often poorly defined. Indeed, some of the issues defy description because of man's limited understanding of precisely what is at stake.

This chapter is an attempt to clarify and outline the expected effects of the physical presence of production equipment and the proposed Trans-Alaska Pipeline (TAP) on vegetation, wildlife, and fishing resources.[1] The last half of the chapter is a discussion of alternative overland routes and how they compare with the TAP-tanker alternative.

Physical and Engineering Challenges

Tundra

The discovery of oil on Alaska's North Slope and the subsequent proposal to transport hot oil by a 48-inch-diameter pipeline nearly 800 miles to the port city of Valdez on Prince William Sound have

[1] These are summarized in detail in app. B.

raised some major engineering challenges. The site of the oil discovery is an Arctic coastal plain. The vegetation cover is tundra, a mat-like cover of grass, lichen, sedges, and other sensitive vegetation over permafrost, or permanently frozen ground. A pipeline built from the North Slope to Valdez would pass through tundra for a sizable portion of its route (map 1) and would certainly have some effect on the tundra vegetation. Moreover, the North Slope is the home of approximately 350,000 caribou who prefer lichens as a year-round food supply and depend on them as a major source of winter food. Since lichens may take from 50 to 150 years to revegetate after disruption, the consequences of tundra disturbance may be severe.[2] Other more serious problems will probably be associated with production equipment, transportation between drilling sites, oil feeder pipelines and gathering facilities, and the production of electric power for the drilling and pumping operations on the North Slope tundra.

Permafrost

A second challenging problem is the effect of the production and transportation of Arctic oil on permafrost. While permafrost can be found throughout most of the state, it is known to exist in almost continuous fashion on the North Slope. Map 2 shows the distribution of permafrost in the area of proposed pipeline routes. Permafrost has been defined as "any ground material that is continually at the freezing point of water or lower, year-round. It can be rocks, gravel, silt, clay, soil, peat, or combinations of these."[3]

Tundra vegetation and permafrost are closely related, since the possibility of thawing depends on the vegetation cover of permafrost. Thawing, in turn, is positively related to water content, which is known to be high in the permafrost on the North Slope. The following describes why production and transportation during the summer may prove to be quite troublesome:

> The top soil of the tundra thaws to a depth of one or two feet, and melted ice and snow, which cannot drain into the frozen subsoil, turn the tundra to bog. If this top layer is removed or sharply compressed, the permafrost is susceptible to thawing, which sometimes can be progressive.[4]

The effect of oil operations on North Slope vegetation and therefore on wildlife—particularly the vast herds of caribou that depend on it—

[2] American Petroleum Institute, *Oil and the Challenge of Alaska* (Washington, D.C., July 1971), p. 4.
[3] Ibid.
[4] Ibid., p. 5.

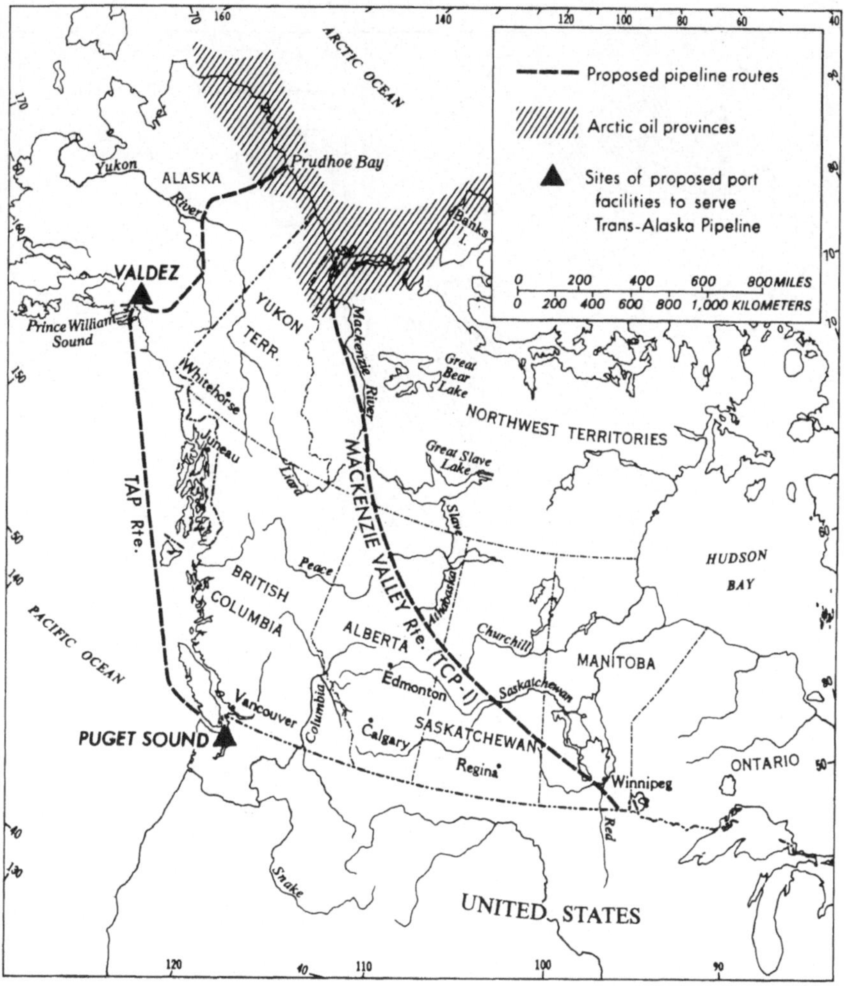

MAP 1. Proposed Arctic oil routes: Trans-Alaska Pipeline and Mackenzie Valley route (TCP-I). *Source*: Charles R. Lewis, "Comments on the Draft Environmental Impact Statement for the Trans-Alaska Pipeline" (submitted to the Department of the Interior, February 17, 1971).

is directly related to summer production activities. Vegetation disruption or heated oil could cause the permafrost to thaw progressively around or under the proposed pipeline, resulting in pipeline displacement. The possibility of pipe displacement has been a major point of disagreement between oil company officials and environmentalists. The former indicate that the pipeline will meet construction specifications

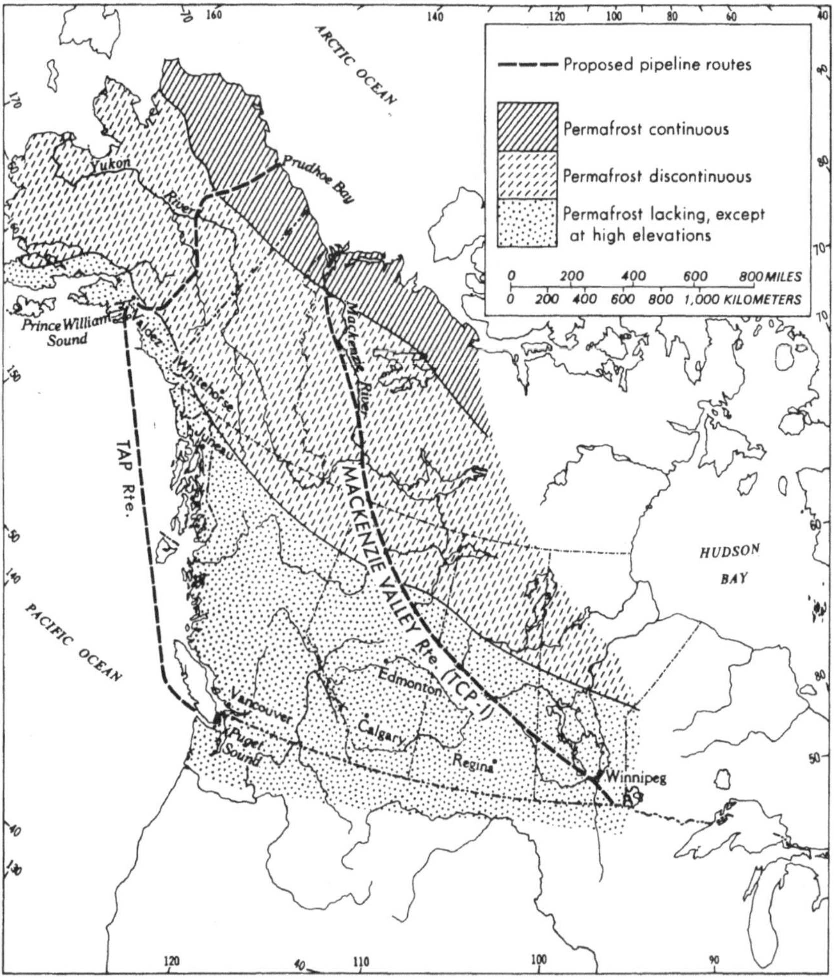

MAP 2. Permafrost areas crossed by proposed Arctic oil pipeline routes. *Source*: See map 1.

that will allow the pipe to withstand stress greater than that of any anticipated permafrost displacement. Many environmentalists, however, have continued to express doubt about whether the specifications will be enforced.

In its *Final Environmental Impact Statement* the Department of the Interior did not give false assurances to environmentalists but outlined the permafrost problem as follows:

Except for the southernmost 25 miles, the proposed Trans-Alaska pipeline route lies in zones of continuous or discontinuous permafrost. . . . Even in the coldest parts of Alaska, a thin layer of ground, the "active layer," thaws every summer and separates the permafrost from the ground surface. . . . Engineering modifications of the environment can cause thawing of permafrost in different ways. The first is by changing the thermal balance at the surface and causing the active layer to thicken from summer heating. Such effects can be caused by gravel pads, damage to the organic surface layer, or disrupted surface drainage . . . but it may take several years before the effects are fully manifest. The second type of modification results from buried hot pipelines.[5]

The department noted that "either effect could cause serious damage to the pipeline."[6]

Construction Methods

There is still some disagreement over how a pipeline should be constructed, most notably over the amount of pipe that would be aboveground and the amount that would be below ground. The Interior Department suggested that as much as 50 percent of the proposed pipeline should be buried in a discontinous fashion, depending on wildlife migratory patterns and the susceptibility of permafrost to thawing.[7]

The amount of sand, gravel, or crushed rock that is necessary for the construction of both the production facilities and the proposed pipeline is of significant concern to those responsible for choosing a construction method. The Interior Department estimated that as much as 80 million cubic yards of sand, gravel, or crushed rock would be used for the pipeline construction, and Alaskan officials estimated that "the gravel necessary for the construction of roads, drilling sites, gathering centers, camps and pipeline installation and other facilities in the Prudhoe Bay Field is . . . about 15.5 million cubic yards."[8]

The oil companies and the environmentalists both agree on the need for stable pipelines to prevent possible breaks, gravel crosswalks to facilitate the migration of wildlife, and gravel to add to the stability of production sites in the tundra of the North Slope. Environmentalists, however, claim that gravel borrows of the magnitude required for construction of the pipeline will severely affect water runoff and

[5] U.S. Department of the Interior, *Final Environmental Impact Statement Proposed Trans-Alaska Pipeline*, vol. 2, *Environmental Setting of the Proposed Trans-Alaska Pipeline* (1972), pp. 7–10.

[6] Ibid., vol. 1, *Introduction and Summary*, p. 97.

[7] Ibid., vol. 2, *Environmental Setting of the Proposed Pipeline*, pp. 40, 42.

[8] State of Alaska, *Comments on the Proposed Trans-Alaska Pipeline*, State of Alaska Report (Fairbanks, July 30, 1971), p. 18.

stream beds in a way that will directly upset the aquatic environment and indirectly upset the food chain of some of the native peoples and various wildlife. On the other hand, while state officials admit that gravel borrows may have such deleterious effects, they maintain that proper regulations and institutional arrangements are being formulated that will result in minimum damage from future gravel borrows (nearly one-third of the 15.5 million cubic yards have already been removed).[9]

Physiographic Considerations

Constructing a pipeline in a north-south direction across Alaska will mean that the engineer must cope with several major physiographic divisions (map 3). TAP would cross two rugged mountain systems, the Rocky Mountain System in the north and the Pacific Mountain System in the south. According to the Interior Department: "In terms of topography and related features, the route crosses terrain analogous to that between San Francisco, California, and eastern Colorado."[10]

The statement of the Department of the Interior includes the following description of the physiography of the proposed route:

[Interior Plains]

The Interior Plains division is represented by the Arctic Coastal Plain, which rises smoothly and imperceptibly from the Arctic Ocean to a maximum of 600 feet at its southern margin.... The Plain is very poorly drained and is therefore very marshy in summer.

[Rocky Mountain System]

The Rocky Mountain system or physiographic division includes the Arctic Foothills, Brooks Range, and the Ambler-Chandalar Ridge and Lowland to the South. The Arctic Foothills ... gradually rise to about 200 feet altitude. ... Rivers from the Brooks Range cross the Foothills in braided course on gravel flats which are locally covered by extensive icings in winter.... Most of the proposed pipeline route through the Arctic Foothills is on the active flood plain or low terraces of the Sagavanirktok River; it does cross some meander necks however. ... The Brooks Range is a wilderness of rugged glaciated east-west ridges that rise to an altitude of 7,000 to 8,000 feet in the northern part of the range and to 4,000 to 6,000 feet in the southern. ... The Ambler-Chandalar Ridge and Lowland consists of east-west lowlands and ridges to 3,000 to 4,500 feet in altitude south of the abrupt front of the Brooks Range.

[9] Ibid., p. 18.
[10] U.S. Department of Interior, *Final Impact Statement*, vol. 2, *Environmental Setting of the Proposed Pipeline*, p. xxiii.

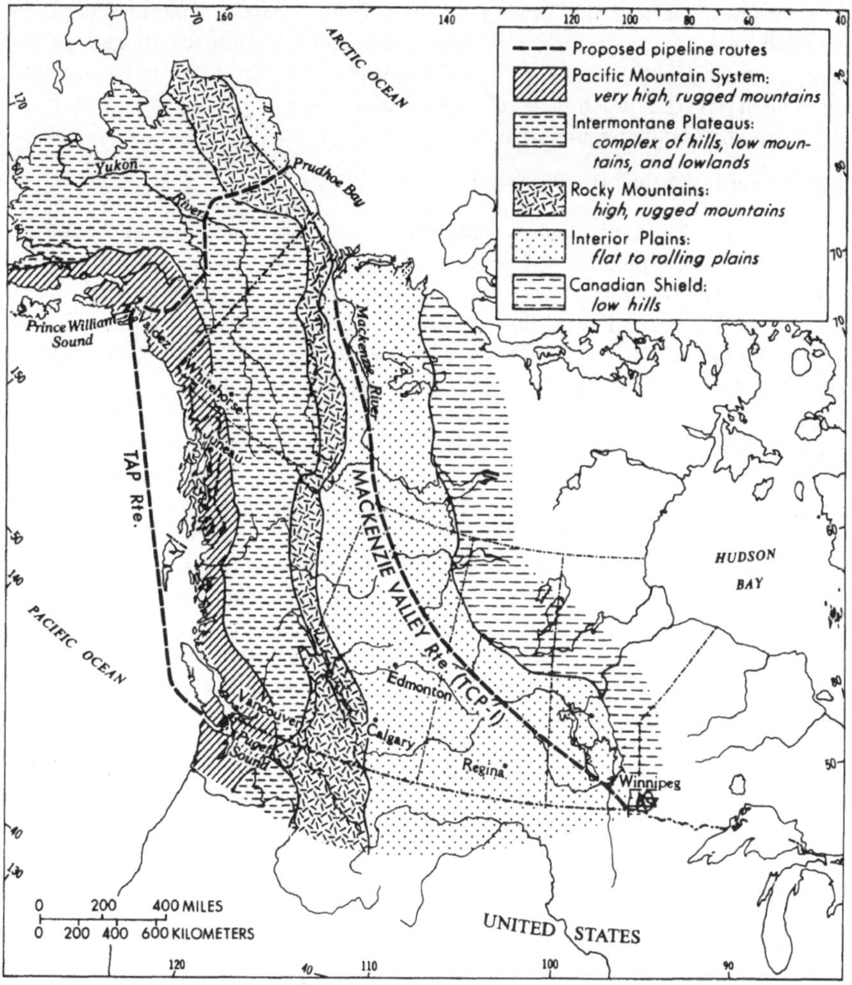

MAP 3. Physiographic regions crossed by proposed Arctic oil pipeline routes. *Source*: See map 1.

[Intermontane Plateaus]

The Intermontane Plateaus include the Kokrine-Hodzana Highlands, Rampart Trough, Yukon-Tanana Upland and Tanana-Kuskokwim Lowland. The Kokrine-Hodzana Highlands consist of even-topped ridges to 4,000 feet altitude with isolated areas of higher mountains. The divide between the Yukon and Koyukuk Rivers passes through the area and small rivers drain in both directions. . . . The Rampart Trough is a depression with gently rolling topography incised from 500 to 2,500 feet below the bordering highlands. The Yukon

River flows in part of the Trough. ... The Yukon-Tanana Upland consists of rounded even-topped ridges with gentle side slopes. Compact rugged mountains surmount the ridges. ... The Tanana-Kuskokwim Lowland is a broad depression on the north side of the Alaska Range. Thaw lakes and thaw sinks are abundant in some areas.

[The Pacific Mountain System]

The Pacific Mountain System consists of the Northern Foothills, Alaska Range, Gulkana Upland, Copper River Basin, and Kenai-Chugach Mountains. The Northern Foothills of the Alaska Range are flat-topped east-west ridges 2,000 to 4,000 feet in altitude, ... with intervening rolling lowlands 700 to 1,500 feet in altitude. ... The Alaska Range consists of parallel, rugged, glaciated groups of extremely rugged snow-capped mountains. ... The Gulkana Upland consists of subdued east-west ridges 3,500 to 5,500 feet in altitude separated by lowlands 2 to 10 miles wide. The lowlands are underlain by glacial deposits. Lakes abound. ... The eastern part of the Copper River Basin is a relatively smooth plain 1,000 to 2,000 feet in altitude trenched by the valleys of the Copper River and its tributaries. ... The Kenai-Chugach Mountains form a rugged barrier along the north coast of the Gulf of Alaska. High parts of the mountains are dominated by very rugged east-west ridges 7,000 to 13,000 feet high with large icefields and glaciers. The coast is deeply indented by fiords and ridges. ... An abrupt front faces the Copper River Basin to the north. ... The area is very active tectonically. ... Faults are probably present but few have been mapped.[11]

The proposed pipeline will cross several major navigable rivers, including the important Yukon River, and as many as 350 additional streams. Some of these crossings may be buried in the river beds, some may be accomplished with the use of gravel and steel culverts, and others may require suspension bridges. Concern about the lack of specification in these construction decisions and about spring ice breakups and ice jams, prompted environmentalists to raise the question of who will make such decisions and how much weight will be given the aquatic resources if the decisions are made by the oil companies themselves.

The environmentalists also are concerned about the problems of containing oil spills and the difficulty of cleanup operations along a pipeline route over such rugged terrain. In prepared comments on the Interior Department's draft environmental impact statement, the U.S. Army Corps of Engineers expressed its concern over the problem of cleanup operations and oil ballast water. It concluded that

[11] Ibid., pp. xxiii–xxxi.

ballast water treatment and precautions against oil spills are closely dependent on water and harbor conditions such as currents, tides, seasonal variations in water volume, etc. ... It should be recognized that there are numerous technical uncertainties in cleaning up marine oil spills and that complete recovery of spilled oil is not likely.[12]

The Corps of Engineers further indicated that it "cannot and will not abandon its regulatory responsibilities under the Rivers and Harbors Act." Under this Act the Corps must issue permits for any hindrances to passage on navigable waters. Furthermore, major rivers have been important transportation routes in Alaska, and the long-range plans of the state indicate that a revival of this mode is likely for the Yukon, Tanana, and Porcupine rivers.[13]

Earthquake Dangers

Perhaps the most notable engineering challenge emphasized by environmentalists is that the route of the lower 70 percent of the proposed pipeline "lies within 25 miles of a recorded epicenter."[14] The most frequently mentioned tanker route to the U.S. West Coast also would pass through areas with a sizable concentration of previous epicenters. Map 4 shows these various epicenter concentrations and three major transcurrent-fault systems that are found along the proposed TAP route. The issues from the environmentalist's viewpoint were well put by Congressman Les Aspin:

The area surrounding the 800 miles over which the pipeline would go has been hit with 23 large earthquakes, with a Richter rating of 6 or more in the last 70 years. Each mile of the pipeline contains twice as much oil as was spilled in Santa Barbara in 1969.[15]

The as yet unresolved controversy is whether it is possible to construct a pipeline that can withstand major earthquakes in this range along the proposed route and—short of such an accomplishment—

[12] U.S. Army Corps of Engineers, Alaska District, *Review Comments—Interior Department Environmental Impact Statement for the Trans-Alaska Pipeline* (February 5, 1971).

[13] This point was made known to me by George Rogers of the University of Alaska in comments on an earlier draft of my analysis.

[14] Charles R. Lewis, "Comments on the Draft Environmental Impact Statement for the Trans-Alaska Pipeline" (submitted to the Department of the Interior, February 17, 1971).

[15] Les Aspin, "It's Not Too Late to Stop the Alaska Pipeline—Maybe," *Not Man Apart*, vol. 1, no. 8 (August 1971).

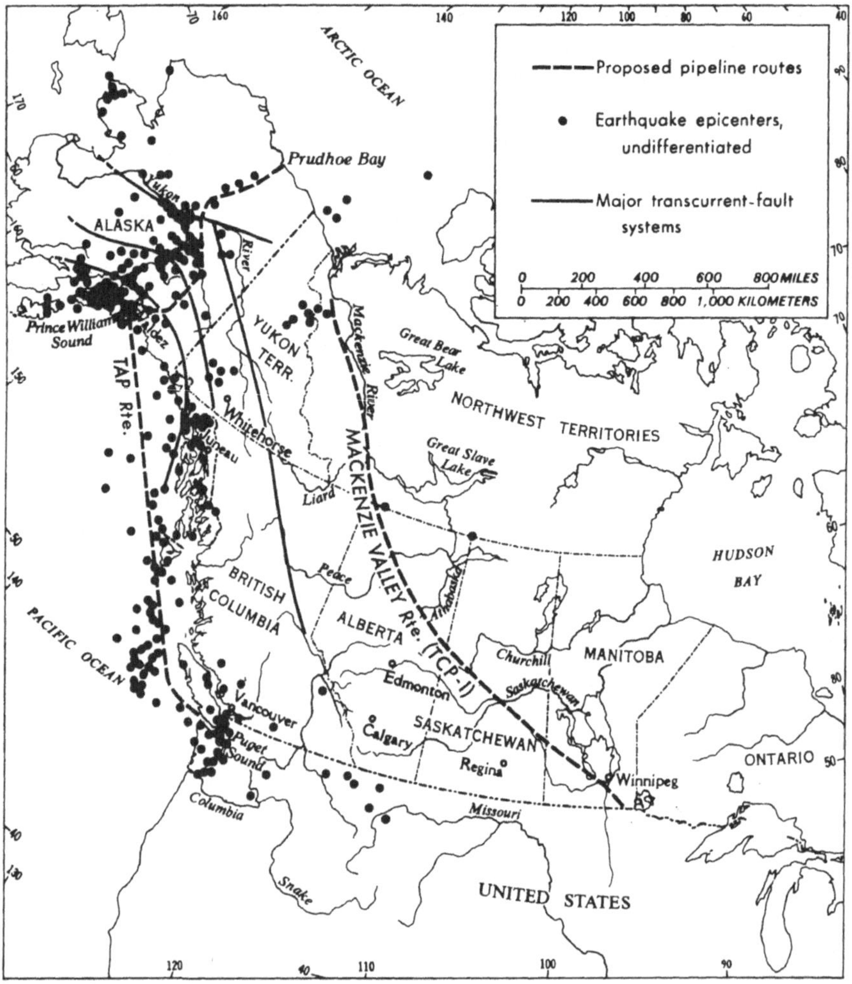

MAP 4. Earthquake zones in the area of proposed Arctic oil pipeline routes. *Source*: See map 1.

whether a system of shutoff valves, retaining dikes, or walls can be constructed to contain or minimize the damage to the wilderness and the many streams and rivers the proposed pipeline will cross. Environmentalists have also raised the issue of possible damage to the fishery resources in Prince William Sound caused by the impact of tidal waves or earthquakes on storage facilities for more than 20 million barrels of oil. This danger has been emphasized by reference to the Good Friday earthquake in 1964, which occurred at the former site of Valdez and had an 8.4 Richter rating, making it the most severe earthquake ever recorded in North America. The counterclaim is that the new site of the terminal city of Valdez is based on solid bedrock and the storage facilities are about 200 feet above sea level. Without trying to evaluate these claims and counterclaims, it is important to note that the earthquake danger is one of great uncertainty and one of the most important points of contention between opposing interests in the pipeline controversy.

The Department of Interior noted the following dangers to the integrity of the proposed pipeline:

> Seismic shaking or surface faulting accompanying a large shock could rupture the pipeline directly or cause failure in the foundation material that could lead to rupture. Furthermore, large earthquakes could trigger landslides and sea waves that could jeopardize the integrity of the pipeline.[16]

The prospect of one or more major earthquakes (magnitude greater than 7 on the Richter scale) along the southern two-thirds of the proposed pipeline route was determined to be "almost a certainty . . . during the lifetime of the pipeline."[17] With the proposed placement of shutoff valves, the maximum expected spill from a pipeline rupture would be 50,000 barrels, or 2.1 million gallons. The damage from a complete pipeline rupture would be greater in the summer than in the winter, since the oil "would be more difficult to contain, recover, or dispose of than a similar loss in the winter both because of its lower viscosity and the probability of being carried on the surface of streams and the sea during the warmer season."[18]

Tanker Traffic

Recent well-publicized accidents involving oceangoing oil tankers have increased the environmentalists' fears about the possibility of such accidents on the high seas off the Pacific Coast and within Prince William and Puget sounds. A large fleet of supertankers would have to cope with hazardous weather conditions while transporting some 2 million barrels of oil per day across Prince William Sound, and estimates run as high as 1 million barrels per day for tankers in Puget Sound. Even if very stringent safety measures were used, the risk would be great.

[16] U.S. Department of Interior, *Final Impact Statement*, vol. 2, *Environmental Setting of the Proposed Pipeline*, p. 11.
[17] Ibid., vol. 1, *Introduction and Summary*, p. 97.
[18] Ibid., vol. 4, *Evaluation of Environmental Impact*, p. 185.

In recent hearings in the Canadian House of Commons chaired by David Anderson,[19] it was pointed out that navigation and crosscurrents would be particularly treacherous in Puget Sound and that Canadian waters are likely to be used if oil is to safely reach the areas that recently were chosen as the sites for oil refineries on the sound. In order to offset the effects of possible collisions and spills, such construction features as (1) double skins and compartmentalization; (2) newly developed high-strength steels; and (3) split rudders, bell flaps, or sea anchors might be investigated to determine their effectiveness in reducing stopping distances and increasing maneuverability. To meet such needs, additional costs of 25 percent for tanker construction were estimated during the hearings.[20]

In the Interior Department impact statement, uncontrollable factors that could create problems for the tanker operation of TAP are described in considerable detail.

The whole coast between Port Valdez and southern California is seismically active. ... The more active portions are Prince William Sound, the Gulf of Alaska, and California. Epicenters of some of the largest historic earthquakes occurred in these areas and the magnitude and frequency of future seismic events are qualitatively predicted to be high.

... The frequency of high waves decreases slightly toward the Gulf of Alaska and south of 50° N there is a steady decrease of frequency of high waves with latitude. Some extremely high wave heights have been reported in the northeastern Pacific. ...

The percentage frequency of restricted visibility (less than 2 nautical miles) reaches a maximum over most of the area in the summer months, although southern California areas peak in the fall or early winter. Coastal fog is so localized that it is not reflected in most ship area summaries. ...

Violent winds frequently occur on the leeward side of the coastal mountain range around the Gulf of Alaska. They occur suddenly and are extremely gusty. ... The same area is marked by restricted visibility 6 to 8 percent of the time in winter months and 6 to 12 percent in summer. ... Prince William Sound is poorly known climatologically. ...

The general vicinity of the Strait of Juan de Fuca and Puget Sound is the site of fogs that form predictably under certain sets of wind and temperature conditions. ... Local effects cause abrupt local variations in fog formation and duration. ... Somewhat similar restricted visibility situations occur in the San Francisco Bay area, one of the differences being the greater persistence

[19] Canadian House of Commons, *Environmental Pollution,* Hearings before the Committee on Environmental Pollution, 28 Parliament, 3 sess. (June 1971).
[20] Ibid., issue no. 21, p. 5.

of fogs there. Restricted visibility is also a characteristic of the Santa Barbara Channel of southern California.[21]

Applying such hazards to a vessel casualty model caused the Department to conclude that "during an average year, about 140,000 barrels would be unintentionally discharged as a result of casualties. This is considered . . . to be an estimate of maximum or worst case casualty discharge volume."[22]

Low-Level Contamination of Prince William Sound

There are two possible sources of low-level contamination of Prince William Sound. The first is any spillage that might be expected from the normal operation of terminal facilities for the loading and unloading of approximately 2 million barrels of oil per day. Second, the ballast water from returning tankers will be treated and injected into the sound at levels of approximately 10 parts per million (ppm). The rate of diffusion of oil depends on the temperature of the water in the sound, which affects the persistency of petroleum and the temporal and areal concentrations of oil residuals.

The range of estimates of normal oil spills given in a recent report is quite wide.[23] For example, if the relatively high spillage rate in Cook Inlet, where both terminal operations and offshore drilling occur simultaneously, were repeated in Prince William Sound, approximately 9.2 million gallons of oil (0.03 percent) would spill into the sound each year. At the present time there is no offshore oil production in Prince William Sound or the Gulf of Alaska. Federal leases may be sold in the 1970s for the Gulf of Alaska. Furthermore, while there have been several conflicting claims about the oil prospects in this area, it is important to note that there was a modest commercial development in the Katalla District in the 1930s. With improved offshore drilling technology and a new major port at Valdez, it seems highly likely that further development will be undertaken.[24] Therefore, problems associated with joint terminal and production operations might arise in the future. At the other extreme, if recent spillage rates for Milford Haven,

[21] U.S. Department of Interior, *Final Impact Statement*, vol. 3, *Environmental Setting between Port Valdez, Alaska, and West Coast Ports* (1972), pp. xviii–xxiv.

[22] Ibid., vol. 4, *Evaluation of Environmental Impact*, p. 474.

[23] State of Alaska, *Comments on the Proposed Trans-Alaska Pipeline*, State of Alaska Report (July 30, 1971).

[24] Point made to me by George Rogers of the University of Alaska in comments on an earlier draft of this analysis.

England, were repeated in Valdez, approximately 12,255 gallons of oil (0.00004 percent) would be spilled annually. The disparity between these estimates is a source of serious concern to the environmentalists and to the fishermen whose livelihood depends on the valuable commercial fishing in Prince William Sound.

In addition to unintentional discharge from terminal operations, the Interior Department considered a number of expected intentional marine consequences from the "normal" operation of the proposed marine transportation system.

> Intentional oil discharge would occur from the ballast treatment facility at the pipeline terminal in Port Valdez and could occur from tank-cleaning and deballasting operations in the open ocean. . . . 'Assuming that the effluent contains a maximum oil content of 10 ppm . . . when the facility is operating at peak capacity, the intentional cumulative discharge of oil from the treatment facility would be 560 gallons . . . per day. . . . [Note: The possibility of double this effluent has been noted.] At the present time the single largest source of intentionally discharged oil on a world-wide basis is from tanker tank cleaning operations.[25]

The department estimated intentional spills along the proposed tanker route at rates of 5,586 gallons per day (assuming Inter-governmental Maritime Consultative Organization standards), 22,680 gallons per day (assuming load-on-top operations), and 62,160 gallons per day (assuming uncontrolled operations). The department indicated that the level of tank-cleaning residues "cannot be accurately predicted" but that it is "likely that a large proportion . . . would go into the shore treatment facility."[26]

Wildlife Migration

Caribou, a joint game resource of Canada and the United States (map 5), are found only in the Arctic. Environmentalists question whether a pipeline can be constructed that will permit the unrestricted migration of the large Arctic caribou herds whose reactions to the physical presence of the pipe, pipeline noise, various overpasses and underpasses, and possible land displacement from melting permafrost cannot be predicted. In general, environmentalists doubt that oil engineers will make decisions in favor of costly expenditures if the only benefit would be the protection of the game resources of the North Slope.

[25] U.S. Department of Interior, *Final Impact Statement*, vol. 4, *Evaluation of Environmental Impact*, pp. 466–68.
[26] Ibid., pp. 468–69.

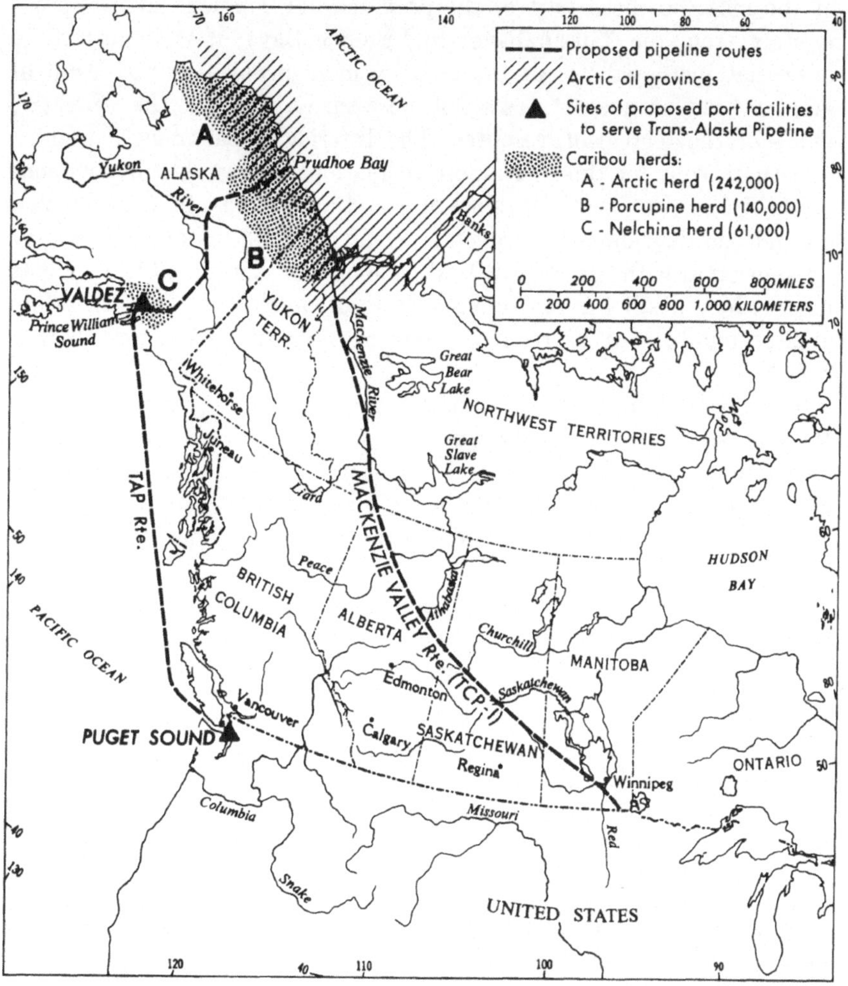

MAP 5. Ranges of caribou herds in the area of proposed Arctic oil pipeline routes. *Sources*: See map 1 and State of Alaska, *Comments on the Proposed Trans-Alaska Pipeline*, State of Alaska Report (Fairbanks, July 30, 1971), p. 18.

Outdoor Recreation

Both the potential and present recreation use of the land and marine resources that would be affected by the proposed pipeline-tanker transportation system are matters of great concern to environmentalists as well as to the Department of the Interior. Given the fragile nature of the most northern part of the proposed development site (approximately from the Yukon River to Prudhoe Bay), it is important to note that increased use of the area by more people (as the result of greater ease of access) may not necessarily increase the economic value of the recreation resource. The Interior Department has given several reasons for this: (1) more recreationists may destroy portions of the fragile environment, since use and access may be concentrated around the development projects; (2) an increase in forest fires may be associated with use; and (3) oil development and possible accidents may reduce aesthetic and other aspects of present use, and therefore greater numbers of users may value the experience less per experience.[27]

The lower portion of the proposed pipeline route (Yukon River to Valdez) has greater recreation use today than the upper portion. Construction, possible accidents, and intrusion that would be aesthetically disturbing would reduce the value by an unknown amount of present recreational experiences that include river boating, canoeing, camping, hunting, and fishing.[28]

More information is available concerning the recreation associated with the marine resources that would be affected by the proposed development. The Department of the Interior noted:

> Prince William Sound and Port Valdez are very important to outdoor recreation because of their outstanding scenic and recreation resources and their proximity to over half of the resident population of the State. Because the area is mostly National Forest, the most common activities—hunting, fishing and photography—take place on USFS lands. Sightseeing and salt water angling also occur on waters of the Sound and contiguous bays, etc. An estimated 47,700 visitor days were spent in the National Forest around the Sound in 1970. Puget Sound provides recreation opportunity for over two million area residents. In 1960 there were an estimated 68 million water-related activity occasions. ... [In] San Francisco Bay ... in 1965 there were an estimated 18 million water-related activity occasions, excluding boat fishing. The southern California coast is of recreational importance because of its proximity to the densely settled metropolitan areas.[29]

It should be noted, however, that the impact of the TAP-related marine transportation, at least on California port areas, would probably

[27] Ibid., pp. 556–58.
[28] Ibid., pp. 558–60.
[29] Ibid., vol. 3, *Environmental Setting between Port Valdez, Alaska, and West Coast Ports*, p. viii.

occur without the proposed Trans-Alaska Pipeline if the alternative source of supply of oil for the California market were other overseas imports.

Some All-Overland Alternative Routes

The controversy surrounding TAP has produced a rather strange environmental anomaly. Contrary to the general belief that environmentalists will oppose all pipelines as an invasion of the wilderness and, therefore, the longer the pipeline the greater the opposition, at least one of the plaintiff organizations, the Environmental Defense Fund, has endorsed the construction of a much longer pipeline from the North Slope through Canada to the U.S. Midwest and East Coast market areas as a superior alternative to TAP.[30] Furthermore, while other environmentalists have not endorsed a Canadian pipeline, many have demanded further study of this alternative since there are some rather glaring environmental differences between these proposed transportation systems.

As with most controversies of this kind, proponents with varying interests tend to exaggerate and overemphasize the differences between alternatives. In this case, opponents and proponents of the Trans-Alaska Pipeline quite naturally emphasize very different environmental and economic points about the Trans-Canadian Pipeline (TCP) alternative. In this discussion I will concentrate on environmental factors, adding to the complexity of the problem by pointing out that there are several alternative routes through Canada. While the routes shown in map 6 are by no means exhaustive, they do provide a basis for considering the various environmental differences among these four Canadian routes and the Trans-Alaska Pipeline.

In its *Final Environmental Impact Statement*, the Department of the Interior presented the alternative pipeline routes in an objective and frank manner.[31] Furthermore, several private and public environmental and engineering studies are being undertaken in Canada, and several government officials have announced that by the end of 1972 enough information will be available to permit a decision to be made

[30] Environmental Defense Fund press release, March 27, 1972.

[31] U.S. Department of the Interior, *Final Impact Statement*, vol. 5, *Alternatives to the Proposed Action.*

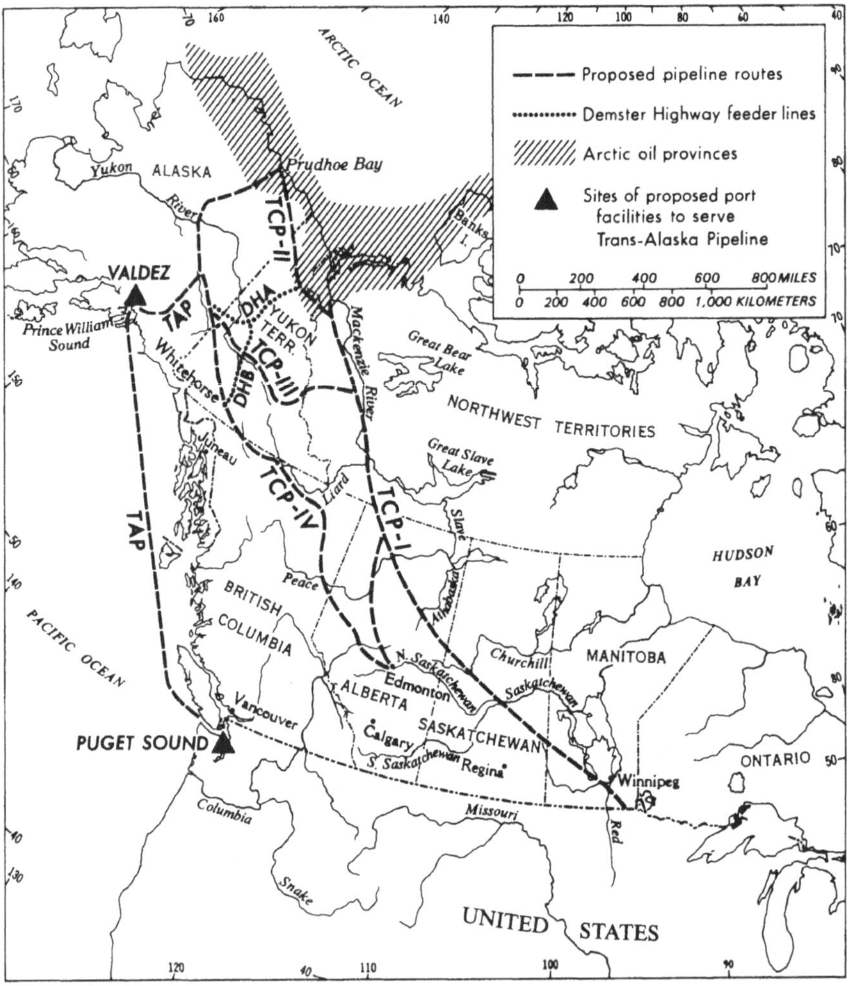

MAP 6. The Trans-Alaska Pipeline route and alternative overland Trans-Canadian Pipeline routes. *Sources*: See map 1 for source of pipeline routes. Digby Hunt, Canadian Ministry of Northern Affairs and Indian Development, supplied the Demster Highway feeder line routes (DHA and DHB).

concerning the application for either a natural gas or crude oil pipeline.[32]

[32] Delays in Canadian approval of a pipeline are likely to take two forms. First, environmental and engineering studies, which are necessary to decide on an application, are not expected to be completed until December 1972. Second, nearly a year would be required to process any application. Some of these steps could occur simultaneously, but if one assumes this will not be the case, as much as a

The Interior Department indicated a further reason for considering these alternative pipeline routes.

One factor involved in viewing potential Trans-Alaska-Canada alternative routes from an environmental impact standpoint is the potential oil and gas resource situation of the entire Arctic and prospective future pipeline development. This is important because the fewer the corridors, pipelines, or other transport systems which might ultimately be used, the lesser would be the injury to nature. ... There are two basic points which need emphasis. ... One is that there are oil and gas resources other than those at Prudhoe Bay in the North American Arctic and the other is that both the oil and gas found in the Prudhoe oil field must be developed since the State of Alaska will not permit "flaring" ...; the gas must either be reinjected or piped to market. ... Under such circumstances there exists the possibilities that 1) since the Prudhoe oil development would eventually fill one 48-inch oil pipeline, the development of Canadian Arctic oil might eventually require another line, and 2) that a gas pipeline would likely be constructed from Prudhoe Bay through Canada to the United States. Thus, the possibility exists for a consolidation of these pipeline developments into a northwestern North America petroleum transport corridor.[33]

The department considered four alternative Trans-Alaska-Canada routes and their respective environmental impacts. Three of these routes are basically the same as three of the routes shown in map 6. In addition, the department considered an offshore pipeline route in the Arctic Ocean. Most analysts, however, including those at the Department of the Interior, do not consider this to be technically feasible at this time. The general environmental issues raised by the department in connection with the overland routes were as follows:

The mountain chains [see map 3] are major obstacles to movement and construction. Less severe, but neverthless important are the topographic hindrances imposed by the intervening hills, plateaus, and uplands. If only

two-year delay (for TCP relative to TAP) could result if all opposition to TAP were removed by January 1973. For a complete review of the position of the Canadian government, as well as a summary of the various environmental and engineering studies, see the submissions made by the Environmental Defense Fund: John F. Dienelt, "Comment Upon Final Environmental Impact Statement Proposed Trans-Alaska Pipeline: The Attitude of the Canadian Government toward Construction of Pipelines through Canada for Delivery of North Slope Oil and Gas" (submitted to the Department of the Interior, May 4, 1972); and John F. Dienelt and H. Lee Watson, "Comment Upon Final Environmental Impact Statement Proposed Trans-Alaska Pipeline: Environmental Information and Studies Developed by Canadian Sources" (submitted to Department of the Interior, May 4, 1972).

[33] U.S. Department of Interior, *Final Impact Statement*, vol. 5, *Alternatives to the Proposed Action*, pp. 125–26.

topography were considered, the most favorable corridor from Prudhoe Bay to Edmonton would follow the coastal plain lowlands and interior plains around the outer margin of the mountains. . . .

On the basis of recorded earthquakes . . . it appears that each of the corridors transverses areas of much lower seismic activity and risk than the proposed trans-Alaska pipeline. In particular, all the new corridors avoid the very high seismic zone in the Prince William Sound area, and those in Northern Canada also avoid the high risk area around Fairbanks. . . .

[The more northern] corridors . . . directly cross or adjoin the 8,900,000 acre Arctic National Wildlife Range, which borders Canada on the Beaufort Sea. . . . This extensive area has been set aside to ensure in perpetuity this expanse of undisturbed Arctic Slope lowland and foothill ecosystem and the unique types of flora and fauna therein.[34]

In addition, the department outlined the river crossing, permafrost concentration, and wildlife issues. Since each Alaskan-Canadian route is very different with respect to these issues, however, general statements of comparison with TAP cannot be made without obfuscation. A final environmental factor is that none of the Alaskan-Canadian routes require the marine transportation system necessary for TAP.

An Environmental Comparison of Alternative Pipeline Routes

One overland route that has been given some attention would closely parallel the Beaufort Sea to Inuvik, more or less parallel the Mackenzie River to Fort Simpson, and then go southeasterly, approaching Edmonton from the west. A right-of-way to U.S. Midwest and West Coast markets in Puget Sound already exists; therefore the terminus in Edmonton is a logical stopping point for most environmental, but certainly not economic, comparisons of the various alternative pipeline transportation systems. I have designated this proposed route as TCP-I. In its final impact statement the Department of the Interior gave the following description of the first section of this route:

This 415-mile corridor is along the river edge of the coastal plain east to the mouth of the Mackenzie River and then south along the narrow benchland on the west side of the Mackenzie Delta. The land is dominantly rolling plains with some low spurs and ridges forming the foothills of the Brooks Range and Richardson Mountains. Weak foundations and unstable slopes are serious engineering problems along a quarter of the corridor, primarily in the eastern part. . . . The unstable ground along the Mackenzie Delta could be bypassed by an alternate corridor 100 miles long across the low ridges and spurs on

[34] Ibid., pp. 127–31.

the east side of the Blow River Basin and in the headwaters of the Porcupine River.[35]

All routes to Fort McPherson would proceed south to Edmonton in the same manner. According to the department:

The corridor . . . trends south 1,240 miles along the Mackenzie Valley and across the Alberta High Plains. Nine major river crossings, extensive areas of ground ice, and local areas of weak bedrock are serious engineering problems.[36]

Another route that has been considered proceeds from the North Slope and traverses both the Brooks Range and the Richardson Mountains, passing through the Old Crow Indian village and the Arctic Wildlife Refuge, then connecting with the former route at a point just south of Inuvik. I have designated this route TCP-II. The northwestern part of this corridor is 465 miles long. "Topography is rugged adjacent to much of the route and engineering problems, primarily concerning weak foundations, are present in about a third of the corridor."[37]

"Trans-Canadian Pipeline" is a misnomer for these routes since they all would pass through portions of Alaska. In fact, the routes I have designated as TCP-III and TCP-IV would be coincident with TAP for approximately two-thirds of its distance in the vicinity of Big Delta and south of Fairbanks. Both routes also would pass through Alaska for nearly as many miles as TAP. The two routes become quite different, however, after entering Canada. The more northern route, TCP-III, follows an existing highway that parallels the Tanana River and turns eastward to the north of Northway following the Ladue River, the Yukon River, and the Macmillan River to Christie Pass, where it then follows the Keele River until it intersects the Mackenzie River to the south of Norman Wells. From this point on to Edmonton, it is identical to TCP-I and TCP-II. Because TCP-III follows so many major waterways it is doubtful that it will have many proponents; in fact, the Interior Department did not even consider this route in its analysis. It is presented here since it is sometimes offered as a possible compromise between the all-Alaskan route and the more northern Mackenzie River routes.

The southernmost Trans-Canadian route, which I have designated TCP-IV, continues to follow existing highways through Northway on Route 1, on to Whitehorse and Watson Lake, where Route 1 becomes Route 97 to Fort St. John. It then follows Routes 2, 34, and 43 through

[35] This corridor is labeled "B" by the Department of the Interior. See U.S. Department of Interior, *Final Impact Statement*, vol. 5, *Alternatives*, pp. 138, 141.

[36] This corridor is labeled "D" by the Department of Interior. Ibid., p. 161.

[37] This route is designated "C" by the Department of the Interior. Ibid., p. 150.

Grande Prairie and on to Edmonton. TCP-IV is sometimes referred to as the Alaska Highway Route. In order to compare TCP-IV more accurately with the other Canadian routes, I have also specified two feeder lines from Inuvik that follow the Demster Highway, which is now being constructed to Dawson. At Dawson a feeder line could flow southwesterly along existing Route 5, picking up TCP-IV north of Northway. I designated this option DHA. Alternatively, the feeder line could follow Routes 3 and 2 southeasterly, picking up TCP-IV in Whitehorse (DHB).

The part of TCP-IV that is different from TAP has been described as follows:

> This corridor extends southeastward for 1,535 miles from Big Delta, Alaska, to Edmonton, Canada ... and generally parallels the Alaska Highway. The corridor traverses nine physiographic regions which contain lowlands, plateaus, and mountains.
>
> The major construction problems encountered would be those associated with more than 20 major river crossings, constricted valleys ..., local areas of permafrost in fine-grain soils, and bentonitic clays in the vicinity of Edmonton.[38]

The distinctive feature of TCP-IV is that it follows motor highways connecting Fairbanks and Edmonton, thus minimizing the effects of constructing multiple corridors—a concern sometimes voiced by environmentalists.

In table 13 environmental problems associated with each of the four TCP routes and with TAP are summarized and compared. This comparison indicates that the two central Trans-Canadian routes are probably inferior to the extreme northern (TCP-I) and extreme southern (TCP-IV) routes. On the basis of physiographic features and the presence of a natural corridor, TCP-I has advantages (as long as the danger of spills into the Mackenzie are not expected). On the basis of possible deleterious effects from permafrost and marine spills, TCP-IV is superior to TCP-I, and it follows an existing man-made corridor. Although a true comparison is somewhat difficult, TCP-IV appears to be environmentally superior to TCP-I.

If a natural gas line alone were considered, a choice between these two alternative Canadian routes would be even more difficult at the present time. If both a gas and a crude oil pipeline were to be built, the possible environmental impact of the Demster Highway connecting line would have to be considered along with TCP-IV. Information on

[38] This corridor is designated "E" by the Department of the Interior. Ibid., p. 172.

this potential feeder line is slight, but its environmental effects would probably be similar to those of TCP-IV since it too would follow a motor highway, although it would certainly be longer and more expensive to construct.

The first point to be made in a comparison of TCP-IV and TAP is that there are no environmental differences between them over the long distance where both follow the same route. TCP-IV is a longer overland route, but because it follows an existing man-made right-of-way the impact of construction activities, the physical presence of a pipeline, and permafrost are lesser problems than they would be otherwise. TCP-IV has further advantages over TAP in that its route avoids the area of intense concentration of seismic activity as well as glaciers and possible avalanches in the Alaska Range. Unlike TAP, TCP-IV does not require terminal operations in Prince William and Puget Sounds, and overland pipelines generally have a better safety record than that of oceangoing tankers. These important environmental differences all favor TCP-IV over TAP. To a lesser extent, the difference in the possible effect of the two routes on the Nelchina Herd and the hazards of weather in the southern portion of TAP also make TCP-IV a more desirable route.

George Rogers, an economist at the University of Alaska, has pointed out two other possible advantages of the TCP-IV route.[39] First, TCP-IV would closely parallel the Haines-Fairbanks military pipeline route for a sizable portion of its route. The engineering and environmental savings would probably be substantial. Second, Rogers pointed out that during World War II a road and pipeline were constructed to transport petroleum from Norman Wells to Whitehorse. This pipeline is not used today, but the road would make it possible to connect Canadian oil provinces into the TCP-IV pipeline at reduced engineering and environmental costs, especially since information concerning construction of the previous project is available.

The Interior Department undertook a similar comparison and concluded that "no single generalized route appears to be superior in all [environmental] respects to any other."[40] The department further concluded, however, that TAP was superior to the Trans-Alaska-Canada routes in only one of six comparisons. The environmental superiority of TAP in that instance was that its impact on the terrestrial abiotic

[39] Letter to Joseph Fisher, President of Resources for the Future, containing comments on a draft of the present study, April 24, 1972.

[40] U.S. Department of Interior, *Final Impact Statement,* vol. 1, *Introduction and Summary,* p. 320.

TABLE 13. A Comparison of Environmental Problems Associated with Alternative Pipeline Routes

Problem	TAP	TCP-I	TCP-II	TCP-III	TCP-IV
Permafrost	Continuous permafrost along northernmost portion for about 25% of route. Discontinuous permafrost for all but southernmost part of route. Ice content could make thawing a major problem in river valley permafrost north of the Brooks Range and in the Yukon-Tanana River valley areas where mixed permafrost is present.	Route traverses nearly twice as much of both continuous and discontinuous permafrost as TAP. Some of this will be Arctic coast tundra permafrost. But water content and more homogeneous topography could make the permafrost problem along this route considerably less than it would be for TAP.	Route passes through the most continuous permafrost but avoids the Arctic tundra permafrost. It does, however, pass through some river valley permafrost in Richardson Mountains.	Route passes through less continuous permafrost than TCP-I and TCP-II but through more discontinuous permafrost than any other route, much of it in river valleys.	Route passes through less continuous and discontinuous permafrost than other TCP routes. Since it follows existing highway system, the hot oil pipeline's impact on and effect from permafrost for most of the route south of Fairbanks will probably be minimal.
Seismic activity	More than ⅔ of TAP route lies within 25 miles of recorded epicenter. Most of the historic seismic activity has been within Fairbanks to Valdez corridor, with major earthquakes at southern terminus, along proposed shipping lane to Puget Sound, and in the Puget Sound area. Possible earthquakes are major concern.	Less than 5% of route is as close as 25 miles from a recorded epicenter—a major advantage frequently cited by environmentalists. But if oil is shipped from Edmonton to Puget Sound by looping the present Trans-Mountain pipeline, this portion of an extended TCP route would pass through seismicly active areas in southwestern portion of British Columbia and northwestern Washington.	Same as TCP-I.	Portion between Yukon River and Big Delta (less than 5% of distance to Edmonton) would pass through same area as TAP and be equally susceptible to seismic activity. But would avoid the main seismicly active portions of TAP and the sea route to Puget Sound. If oil flows along the Trans-Mountain route after reaching Edmonton, route would be susceptible to seismic activity in Puget Sound area.	Similar to TCP-III.
Marine spills	Entire TAP throughput would pass through terminal facilities in Prince William Sound (less a small amount that might be consumed in Fairbanks). Probable oil spills during *normal* conditions have been major concern of environmentalists, as well as possible accidents from collisions, ice, and tidal waves.	Possible damage to Arctic Ocean because route crosses many rivers flowing into ocean in a remote area and passes very close to the coast. There is concern over possibility of serious damage a leak or pipeline break could cause.	Marine pollution not a primary problem to Edmonton. If oil were sent to Seattle and then to Los Angeles or San Francisco via ocean-going tankers (an unlikely economic decision), possible pollution in Puget Sound would be a major concern.	Same TCP-II.	Same as TCP-II.

Physiography	Route passes through two major mountain ranges. Brooks Range is traversed in the north following a generally natural pass. South of Fairbanks two high peaks and avalanche and ice-slide potential in Alaska Range could present hazards. Also, land between these two ranges is complex of hills, low mountains, and lowlands.	TCP-I is superior route from standpoint of naturally homogeneous topography, since it is made up almost entirely of flat to rolling plains and eastern foothills.	TCP-II cuts across Brooks Range and Richardson Mountains in a manner presenting serious engineering and environmental challenges before picking up more natural corridor provided by the Mackenzie River.	Route crosses the Brooks Range in more advantageous manner than TCP-II. However, a significant portion must pass through complex of hills, low mountains, and lowlands and through the Mackenzie Mountains.	Route passes through physiographic features similar to TCP-III but does so following an existing man-made and natural corridor.
Wildlife migration	Major animal whose migration might be affected by pipelines is the caribou. TAP route passes between two main Arctic herds and one in southern Alaska. Porcupine Herd most likely to be affected, but effect is unknown. While overpasses and underpasses are almost certain to be constructed, possible delays in crossing will likely affect calving and grazing site selections, thus reducing herd size.	Route would also affect Porcupine Herd in unknown manner insofar as possible delays could affect the optimal size of the herd.	Route will probably have a more serious effect on the Porcupine Herd since it would just about equally bisect the herd's migration area. Route also would cross the Arctic Wildlife Refuge and pass through Old Crow, a Canadian Indian village.	Route similar to TAP but avoids the Nelchina Herd in the Alaskan Range.	Same as TCP-III.
Climate & glacial activity	Rainfall heaviest along any of proposed routes (about 40 in./year) in area around Valdez. Heavy snow and ice and potential avalanches and glaciers most likely to be more severe problems in Alaska Range or southern part of TAP.	Rainfall is lightest along northern part of this route (6–12 in.). However, "dry snows," or sand-and-snow dust storms, would be more severe for TCP-I, making driving and visibility virtually impossible. Harsh winters in Upper Mackenzie.	TCP-II passes through part of Brooks Range most likely to be affected by avalanches and glaciers, although these are not expected to be as prevalent as in that portion of TAP that passes through the Alaska Range.	Winter climate of portion of route that passes through central Yukon Territory normally expected to be least severe. Rainfall in northern British Columbia and area of Great Slave Lake is about 20 in./year. Brooks Range hazards and problems are same as for TAP ,but hazards of Alaska Range and Valdez winters are avoided.	Similar to TCP-III.

SOURCES: This table is based on several sources, listed in the order of their importance: (1) Charles R. Lewis, "Comments on Draft Environmental Impact Statement For The Trans-Alaska Pipeline," submitted to the Department of the Interior, February 17, 1971; (2) R. A. Rice, *The North American Arctic Coast Energy Transport and Resource Distribution Problems* (Carnegie-Mellon University, Transportation Research Institute, February, 1971); (3) State of Alaska, *Comments on the Proposed Trans-Alaska Pipeline*, State of Alaska Report (Fairbanks: July 30, 1971); (4) American Petroleum Institute, *Oil and the Challenge of Alaska* (Washington, D.C., June 1971); and (5) U.S. Department of the Interior, *Final Environmental Impact Statement: Proposed Trans-Alaska Pipeline*, vols. 1–5 (1972).

(nonliving) environment would be less than that of the all-overland routes, but this would not be so if a natural gas pipeline were to be built in the future across Canada. In each of the other categories the department conceded that one of the Trans-Alaska-Canada routes was superior to TAP. From the standpoint of the impact on the overall biotic environment, the route I labeled TCP-I was superior. From the standpoint of the "unavoidable impacts upon . . . recreation, aesthetics, wilderness, communities, and native culture and subsistence," the TCP-I route was superior. According to the department, TCP-II and TCP-IV would probably have the least impact on the marine environment. From the standpoint of risk, or threatened environmental impact, the conclusion was that the Trans-Alaska-Canada routes were superior to TAP for both the terrestrial and marine environments.[41]

The Interior Department pointed out that assigning weights to these differences might permit more extensive analysis. The selection of weights, however, was thought to be indeterminate. In a comment on the department's analysis, A. Myrick Freeman III assigned various weights to these environmental measures, tested various parameters for sensitivity, and concluded:

> Sample calculations show that the relative standings of the two projects [TAP versus TCP-I] are most sensitive to the weight given to tanker transport miles, since this is the largest one in the TAP score. If tanker miles are deemed more hazardous to the environment than ordinary land pipeline miles (and this is the informed opinion of experts), the balance is shifted so far toward the Canadian route that further calculations are not necessary. . . . Under some conditions, TAP appears to be preferred to the Canadian route. In order to reach this conclusion, the weight placed on tanker miles must be reduced to about .5, seismic effects must be given near zero weight, and higher weights must be placed on continuous and discontinuous permafrost zones.[42]

Since Freeman finds little to justify these weights, he concludes that TAP is environmentally inferior. Most analysts concur. Further, there seems to be no disagreement about the absolute superiority of a joint gas-oil corridor through Canada over a system with two separate corridors.

If definitive conclusions concerning the Trans-Canadian alternatives and TAP are to be drawn, several additional points must be considered. First, as mentioned earlier, a pumping facility is being installed at

[41] Ibid., pp. 320–22.
[42] A. Myrick Freeman III, "An Evaluation of the Final Environmental Impact Statement Proposed Trans-Alaska Pipeline" (submitted to the Department of the Interior, May 4, 1972).

Edmonton with a capacity to send 2 MMb/d to the Midwest. Since the amount of Canadian oil that is permitted to enter the United States east of the Rockies is restricted,[43] this pipeline will have excess capacity to handle a large portion of oil from the North Slope. On environmental grounds, a possible break or leak for a larger throughput would certainly be worse than for the present throughput. Such an eventuality is unlikely, however, and thus the extra environmental consequences of an expanded Interprovincial Pipeline are almost negligible.

A second fact that must be considered is the possible effect on Puget Sound if a Trans-Canadian rather than a Trans-Alaska pipeline is built. If oil from either the North Slope or Alberta is diverted to the Seattle and Vancouver market areas overland to meet the areas' growing needs, it would certainly be counted as an additional environmental advantage of a Trans-Canadian route, since it would minimize the oil tanker traffic in Puget Sound. On the other hand, if tankers are used on Puget Sound as a means of carrying Alaskan oil indirectly to the California market areas, the environmental impact on Puget Sound could be worse. Similarly, if TAP were built and Puget Sound tanker traffic were used to displace the present Canadian oil that supplies the Seattle market area, or if Puget Sound were used as a terminal to supply some midwestern markets by an overland route, the possible damage from tanker and terminal spills would increase in the sound.

Having made these qualifications, I conclude that on environmental grounds TCP-I or TCP-IV would probably be superior alternatives to supply the United States with North Slope oil than either TAP or other Alaskan-Canadian alternatives. The only possible exception to this might occur if Puget Sound were used as a major exporting terminal for either the TCP-I or TCP-IV alternative. This is unlikely, however, as I will show in the next chapter when I consider these routes on economic grounds. In addition, a comparison of the real cost to the United States and Canada of constructing these alternatives is necessary before one can decide whether the environmental differences discussed here are sufficiently large to warrant selecting a Trans-Canadian alternative over TAP.

[43] See app. A for further discussion of this point.

3

An Economic Analysis of Alternative Overland Pipeline Routes Through Canada

In my analysis of the economic benefits and costs of utilizing the proposed Trans-Alaska Pipeline (TAP) and marine tankers to transport North Slope oil to U.S. West Coast markets, I attributed all the benefits derived from production of this domestic oil to TAP. If alternative pipeline routes or other transportation systems are compared with TAP, however, it is incorrect to attribute all the benefits of North Slope oil production to a single transportation alternative.

The purpose of this chapter is to expand the earlier economic analysis by relaxing several important assumptions so that alternative transportation systems and sources of supply can be systematically considered. In chapter 2 several environmental advantages were discussed that could be attributed to a completely overland pipeline route through Alaska and Canada but not to the TAP mixed land and water transportation system. Presumably TAP is economically superior. Therefore, it is important to quantify the economic costs and benefits of the environmentally superior transportation systems so that both the environmental and economic attributes can be weighed in order to derive the socially optimal pipeline route. This can be accomplished by quantifying the net social benefits of a completely overland pipeline route and subtracting the net social benefits of TAP. It is then possible to compare any loss in social benefits from constructing an overland pipeline route through Canada with the environmental advantages of such routes. This comparison is the logical way to analyze alternatives to determine the amount the nation must be willing to pay to lessen any deleterious effect on the environment from the transportation of North Slope oil to the lower forty-eight states.

The various components of benefits and costs should be divided into two categories—those that are sensitive to the transportation system that is selected and those that are not.

1. Components of benefits and costs that depend on the transportation system and domestic market that is selected:
    (a) Environmental effects of transportation.
    (b) Costs of alternative sources of supply.
    (c) Opportunity cost of alternative routes.
    (d) Equity effects in various U.S. markets.
    (e) Supply and demand imbalances.
2. Components of benefits and costs that are independent of the transportation system that is selected:
    (a) National security benefits (there may be some difference in transportation systems in terms of tactical vulnerability, but I consider this a secondary factor that doubtless is an advantage of the all-land Trans-Canadian alternatives).[1]
    (b) Future value of petroleum as a source of energy (assuming both systems would be built at the same time).
    (c) Environmental effects caused by production.

In this chapter, I will presume that oil on the North Slope will be produced and avoid further consideration of the second group of components since they will be generally similar, regardless of the transportation system that is selected. Furthermore, the detailed analysis of the environmental effects of alternative pipeline routes in the previous chapter covered the environmental effects of transportation. Having concluded (1) that when all factors were considered the Trans-Alaska Highway route (TCP-IV) was environmentally superior on net to the TAP system, and (2) that the northernmost Mackenzie River route (TCP-I) also was probably environmentally superior to TAP, I will concentrate on the cost of alternative sources of supply and the opportunity cost of alternative markets for these two routes. I will postpone discussion of the equity effects and supply and demand imbalances until succeeding chapters, where I consider various institutional and

[1] The Office of Emergency Preparedness concluded that surplus supplies on the West Coast would mean that the TAP system was inferior to a Trans-Canadian pipeline from a national security standpoint until 1980, after which time the two alternatives would be largely equal. See U.S. Department of the Interior, *An Analysis of the Economic and Security Aspects of the Trans-Alaska Pipeline*, vol. 2, *Supporting Analyses*, app. M (December 1971).

political factors and further relax the assumptions made in chapter 1.

Here, I will calculate the benefit-cost ratios for the Midwest and East Coast markets of the United States, when (1) a long-run view of the U.S. oil industry is taken, and (2) it is assumed that variations in domestic costs are unimportant between District V and the eastern regions of the country and that foreign oil is the relevant alternative source of supply in all oil deficit regions of the United States. Subsequently, I will deal with what may be a more realistic case in the present and near future and will consider the situation in which a combination of some higher-cost domestic crude, as well as foreign crude, will be displaced by North Slope crude in each domestic market. Finally, I will make some additional adjustments for quality differences in crude oil in specific U.S. markets.

I will use a computer simulation program similar to that in chapter 1 to test the sensitivity of several policy variables to variations in some important parameters and assumptions. The policy variables are net social benefits (economic efficiency), state tax revenues, and corporate profits.

The Cost of the Trans-Canadian Pipeline (TCP)

Any interested party who attempts to keep track of the various public estimates of the cost of a Trans-Canadian pipeline will probably have a difficult time of it. Estimates of construction costs ranging from $1 billion to $7 billion have appeared in the statements and reports of various oil companies and public officials in both Canada and the United States. There are several reasons for this apparent confusion, besides any possible gamesmanship that might be involved in such announcements.

First, since there is not one pipeline route but several, cost differences are likely, especially for the longest—the Alaska Highway route (TCP-IV). Oil might be transported from the North Slope to Edmonton, where it could then be shipped to higher-priced Midwest markets or sent to the Puget Sound area of the West Coast. The Midwest market is superior on both economic and financial grounds, and it would therefore be chosen in the absence of any governmental intervention. Both the West Coast and the Midwest are now supplied with oil from Edmonton by pipelines that could be looped, thus tending to reduce the capital investment necessary for any new pipeline, as well as most of the environmental consequences.

Part of the confusion over the capital cost estimates of the TCP systems is because it is sometimes assumed that the existing routes will be looped and sometimes that a completely new pipeline will be built. Looping is the most likely and the least costly. As indicated earlier, the Interprovincial Pipeline, which carries oil from Edmonton to the U.S. Midwest and to eastern Canada, is being expanded to a pumping capacity of 2 million barrels per day, and in some sections the first 48-inch-diameter pipe ever laid in North America is being used. This throughput level exceeds the amount that Albertan oil is likely to supply, given present restrictions on importing Canadian crude oil in the eastern portion of the United States. The pipeline could be utilized, therefore, to carry a large portion of North Slope crude oil to the Midwest markets at much lower capital costs than a new system.[2]

A second point of confusion over the capital cost of a Canadian route is that it is sometimes assumed that the oil produced on the North Slope will be piped all the way to the East Coast, thus increasing the capital costs of the system. The important fact about such an extension of the route beyond the Chicago market area is that profits and wellhead prices (and therefore taxes to the state) would be unaffected, since the price differences in these two U.S. market areas equal present and expected future transportation cost differences. Depending on which source of supply is displaced by North Slope crude, however, these two market areas each may yield different net social benefits for the nation.

A final source of confusion over the cost of the TCP system is the assumption that the crude oil pipeline through Canada will be coincident with a natural gas pipeline for at least part of its route. Statements of system capital costs based on this assumption generally al-

[2] One of the reasons that the pumping capacity of the Interprovincial Pipeline is being expanded may be completely unrelated to North Slope transportation problems. In 1971 the U.S. government restricted the amount of Canadian crude that may enter the United States to 395,000 barrels per day. The stated reason for this restriction was that the eastern provinces of Canada were becoming overly dependent on relatively insecure foreign sources of oil (mainly Venezuela) and that the U.S. government must consider these eastern Canadian provinces in the U.S. sphere of national security. Accordingly, it was felt that a decision to stop the increase in the flow of Canadian oil into the United States would result in an expanded delivery system within Canada to supply its own eastern provinces with more secure North American crudes. In the interest of fairness it should be noted further that Canadian crude oil also began to threaten the market control of the domestic producers in District III in the late 1960s, which certainly contributed to the political reality of this decision. Building an expanded throughput capacity might restore Canada's overland exemption under the quota system of the Mandatory Oil Import Quota Program. If the Interprovincial Pipeline is being expanded for these reasons, this expansion would be only indirectly related to Arctic oil transportation needs.

most double the cost of a single overland oil pipeline. I will consider each of these situations with varying degrees of precision.

Cost Determination

In table 14, I present a set of calculations for estimating the capital costs of a Trans-Canadian pipeline through the Mackenzie Valley,

TABLE 14. Capital Cost Estimates for the First Segment of a Trans-Canadian Oil Pipeline from the North Slope to Edmonton

(billions of 1971 dollars)

	Capital cost—Direct route to Edmonton (TCP-I, II)		
	Low	Medium	High[a]
Item	$2.20 bil.	$2.38 bil.	$2.5 bil.
1,800 miles at $1.2 mil./mi.	1.98	2.16[b]	2.16
Other capital investment	0.22[c]	0.22[c]	0.34[d]
Total	2.20	2.38	2.50

[a] In a recent analysis, Nehring used a total system cost of $5.3 billion in 1974 dollars. Converting this to 1971 dollars at a 10 percent discount rate makes his total cost estimate to Chicago $4 billion, and splitting this at 66 percent for the North Slope to Edmonton segment results in his comparable estimate being equal to $2.63 billion in 1971 dollars. Although this estimate is slightly higher than those shown in this table, each of the estimates in the table is adjusted upward by 10 percent and 20 percent for sensitivity in the analysis. Therefore, Nehring's case is well within the range of sensitivity used for the North slope to Edmonton segment, namely $2 billion to $3.3 billion, which is used in my analysis for the route labeled TCP-I. See: Richard Nehring, *Future Developments of Arctic Oil and Gas: An Analysis of the Economic Implications of the Possibilities and Alternatives*, Department of the Interior, Office of Economic Analysis (May 10, 1972).

[b] The cost per mile for the Arctic oil pipeline was based on an estimate used in Richard A. Rice, *The North American Arctic Coast Energy Transport and Resource Distribution Problems* (Carnegie-Mellon University, Transportation Research Institute, February 1971). The high and medium estimates are based on the same costs per mile as the TAP route. The low estimate utilizes an average of the TAP cost per mile and the cost per mile used by Rice for a Trans-Canadian gas pipeline of the same size and along a similar route. Since the topography is more favorable along the proposed Canadian route, this low cost figure may be the most accurate for comparison with TAP.

[c] Based on extrapolations from total capital cost figures utilized by Walter B. Parker in *A Comparison of Prudhoe Oil Costs via Valdez or via a Mid-Canada Pipeline*, Federal Field Committee for Development Planning in Alaska (Anchorage, March 1971). The medium cost figure was determined by subtracting $120 million (100 miles × $1.2 million miles) from his estimate of $2.5 billion in 1971 prices for 1,900 miles of pipeline along similar routes for part of the Prudhoe to Edmonton pipeline.

[d] This is a "guesstimate" to round off the mileage costs to a frequently quoted figure.

using approximately the same costs per mile that I used for TAP. This route avoids most of the expenditures for terminal facilities and all the expenditures for marine facilities that are included in the TAP estimates. In table 14 I considered estimates of the capital cost of a crude oil pipeline from the North Slope to Edmonton ranging between $2 billion and $2.75 billion in 1971 U.S. dollars. The computations for cost per barrel for these capital cost estimates and the three discount rates (8, 10, and 12 percent) used in the previous chapter are shown in table 15.

Note that table 15 also includes a new set of computations that were made to account for the possible two-year delay that might be required for a Canadian pipeline. These computations show that when a two-year delay with an accelerated throughput schedule (2 million barrels per day) for a TCP is compared with a no-delay Alyeska schedule, the average annualized capital cost per barrel is nearly equal. That is, for the discount rates considered, a two-year construction delay followed by an accelerated throughput schedule for the TCP just offsets (actually it saves about 1¢ per barrel) any gain in the present value of costs per barrel that the slower production schedule (implicit in the Alyeska rate of throughput) starting two years earlier would have made.

The second component of costs per barrel is the operating cost per barrel of crude oil. The three major North Slope companies estimated

TABLE 15. Per Barrel Capital Cost for Direct Trans-Canadian Pipeline to Edmonton from the North Slope

(dollars per barrel)

Discount rate	Capital cost ($ billion, 1971)			
	2.00	2.25	2.50	2.75
	Alyeska throughput			
8%	0.388	0.436	0.485	0.533
10%	0.495	0.557	0.619	0.681
12%	0.622	0.699	0.777	0.855
	Accelerated throughput			
8%	0.323	0.364	0.404	0.445
10%	0.402	0.452	0.502	0.552
12%	0.491	0.552	0.613	0.675
	Accelerated throughput after a two-year delay			
8%	0.377	0.424	0.471	0.519
10%	0.486	0.547	0.608	0.668
12%	0.616	0.693	0.770	0.846

this operating cost for the North Slope-Edmonton-Chicago system, operating at 2 million barrels per day (730 million barrels per year), to be approximately 30¢ per barrel,[3] or $216.9 million per year. In table 16 this operating cost is broken down into costs of 17.4¢, 12.6¢, and 6.5¢ per barrel for the North Slope to Edmonton, Edmonton to Chicago, and Edmonton to Puget Sound, respectively. In table 17 the estimates of total cost per barrel for the North Slope to Edmonton segment are shown for the various capital costs, throughput schedules, and discount rates and an operating cost of 17.4¢ per barrel.

TABLE 16. Operating Costs for Full-Capacity Throughputs

	North Slope to Edmonton (48″)	Edmonton to Chicago (48″)	Edmonton to Puget Sound (36″)
Expenses for personnel, materials, maintenance, service, administration, oil measurement losses, and fuel (*Can. $*)	35,300	41,220	12,390
Depreciation (*Can. $*)	42,400	14,830	8,190
Total annual operating costs (*Can. $*)	77,700	56,050	20,580
Maximum annual throughput (*mil. bbl.*)	547.5	547.5	292
Operating costs/bbl at maximum throughput (*Can. ¢*)	14.2	10.2	7
Linear increase in costs/bbl to a throughput level of 2 MMb/d (*Can. ¢*)[a]	18.9	13.6	n.a.
	32.5		
Conversion to U.S. dollars, Can. $1 = U.S. 1971 $0.93 (*U.S. ¢*)	17.4	12.6	6.5
	30		

SOURCE: Canadian Bechtel, Ltd., *Mackenzie Valley Pipeline Research Limited, Preliminary Report* (June 1970).

NOTE: n.a. = not applicable. Monetary costs of property taxes and interest have not been included above.

[a] If costs are expected to increase linearly when throughput is increased from 1.5 MMb/d, the assumption is that savings from economies of scale are offset by additional pumping costs. This assumption has been utilized generally in the estimates found in the various Arctic pipeline costs reports prepared for or by the oil industry.

The capital cost estimates per barrel for the Edmonton to Chicago leg depend on the assumption that is made about the costs of looping

[3] See app. B of the memorandum of Atlantic Richfield Company, "Re Trans-Canada Alternative Route" (submitted to the Department of the Interior, September 10, 1971).

TABLE 17. Total Cost per Barrel for a Direct Trans-Canadian Pipeline to Edmonton from the North Slope with Constant Operating Costs

(*dollars per barrel*)

Discount rate	Capital cost ($ billion, 1971)			
	2.00	2.25	2.50	2.75
	Alyeska throughput production schedule			
8%	0.562	0.610	0.659	0.707
10%	0.669	0.731	0.793	0.855
12%	0.796	0.873	0.951	1.029
	Accelerated throughput production schedule			
8%	0.497	0.538	0.578	0.619
10%	0.576	0.626	0.676	0.726
12%	0.665	0.726	0.787	0.849
	Accelerated throughput after a two-year delay production schedule			
8%	0.551	0.598	0.645	0.693
10%	0.660	0.721	0.782	0.842
12%	0.790	0.867	0.944	1.020

NOTE: These calculations are based on a per barrel operating cost of 17.4¢

or using existing excess capacity on the Interprovincial Pipeline and may be as small as $0.1 billion.[4] On the other hand, frequently cited figures for a TCP to Chicago range from $3 billion to $3.5 billion—estimates that are nearly ten times (or $1 billion) greater for this route. To increase the estimates first cited in various early oil company reports,[5] capital costs of $0.6 billion, $0.8 billion, and $1 billion will be analyzed.[6]

The capital costs per barrel are presented in table 18 for each capital

[4] See "Interprovincial Adds a Quarter-Million Horsepower," *Oil and Gas Journal*, October 18, 1971, pp. 68–72.

[5] See Atlantic Richfield, "Re Trans-Canada Alternative Route."

[6] Converting the estimate made by Nehring into 1971 dollars in the manner described in fn. a, table 14, shows his estimate to be slightly higher for this segment, namely $1.35 billion. Note, however, that while his total capital costs in 1971 dollars are about $0.25 billion greater ($3.98 billion minus $3.75 billion) than the high case shown in table 20, it is well below the capital cost case of $4.5 billion, which is used later when the results in table 20 are further tested for sensitivity and adjusted upward by 10 percent and 20 percent. See Richard Nehring, *Future Developments of Arctic Oil and Gas: An Analysis of the Economic Implications of the Possibilities and Alternatives*, Department of the Interior, Office of Economic Analysis (May 10, 1972). He estimates the costs in 1974 to be $5.3 billion, which equals $3.98 billion in 1971 dollars at a 10 percent discount rate.

TABLE 18. Capital Cost per Barrel when the Full Throughput Is Piped from Edmonton to Chicago

(dollars per barrel)

Discount rate	Capital cost ($ billion, 1971)		
	0.60	0.80	1.00
	Alyeska throughput production schedule		
8%	0.116	0.155	0.194
10%	0.148	0.198	0.247
12%	0.186	0.249	0.311
	Accelerated throughput production schedule		
8%	0.097	0.129	0.162
10%	0.121	0.161	0.201
12%	0.147	0.196	0.245
	Accelerated throughput after a two-year delay production schedule		
8%	0.113	0.151	0.189
10%	0.146	0.194	0.243
12%	0.185	0.246	0.308

NOTE: These calculations are based on a per barrel operating cost of 12.6¢.

cost, throughput, and discount rate. By using operating costs of 12.6¢ per barrel for the Edmonton to Chicago segment, the total costs per barrel can be estimated. These are shown in table 19. Finally, one can add the total cost estimates per barrel for the North Slope to Edmonton (table 17) and Edmonton to Chicago (table 19) segments and determine the total North Slope to Chicago transportation cost estimates per barrel (table 20).

Note that the extra costs per barrel caused by a two-year delay just offset the extra costs per barrel associated with the slower Alyeska production schedule. There are two reasons to suppose that the full schedule of 2 million barrels per day might begin in the first full year of operation of a TCP system. One important difference between the two transportation systems is that the principal market the TAP system would serve (the U.S. West Coast) consumed approximately 2 million barrels of oil per day in 1970. On the other hand, the principal market the TCP system would serve (the U.S. Midwest and East Coast) consumed approximately 10 million barrels of oil per day in 1970. If North Slope oil production rates prove to be lower than the accelerated throughput schedule of 2 million barrels per day, the possibility of Canadian Arctic or Mackenzie Valley oil making up the difference

TABLE 19. Total Cost per Barrel when the Full Throughput is Piped from Edmonton to Chicago with Constant Operating Costs

(dollars per barrel)

Discount rate	Capital cost ($ billion, 1971)		
	0.60	0.80	1.00
	Alyeska throughput production schedule		
8%	0.242	0.281	0.320
10%	0.274	0.324	0.373
12%	0.312	0.375	0.437
	Accelerated throughput production schedule		
8%	0.223	0.255	0.288
10%	0.247	0.287	0.327
12%	0.273	0.322	0.371
	Accelerated throughput after a two-year delay production schedule		
8%	0.239	0.277	0.315
10%	0.272	0.320	0.369
12%	0.311	0.372	0.434

NOTE: These calculations are based on a per barrel operating cost of 12.6¢.

is greater for the TCP alternative than for TAP. Delays will provide additional time for further exploration in both Canada and Alaska.

A second important comparison that should be made is between the estimated transportation costs per barrel in table 20 for TCP-I and similar cost estimates per barrel for a TAP-tanker system to the Los Angeles market (table 8). This comparison indicates that throughput without delay for TCP-I has lower costs per barrel than TAP system estimates using the Alyeska schedules for comparable cases.[7] Furthermore, the range of estimates for the Alyeska schedule in both tables begins at about 80¢ per barrel and ranges to about $1.26 per barrel for the TAP system and $1.46 per barrel for the TCP-I system. This difference may become unimportant when one considers such additional factors as (1) the possible overestimate of the capital costs between Edmonton and Chicago, (2) the present higher prices on the East Coast and in the Midwest compared with those on the West Coast, (3) the likelihood of the more rapid full production schedule (even if delayed two years), and (4) environmental differences between the two systems.

[7] In this discussion and elsewhere in the text, "comparable" means high-cost cases are considered with each other, medium with medium, and low with low. In table 22, some of these cost comparisons are presented jointly.

TABLE 20. Total Cost per Barrel from the North Slope via a Mackenzie Valley Crude Oil Pipeline to Edmonton and then to Chicago (TCP-I)

(dollars per barrel)

Discount rate	Estimated capital cost ($ billion, 1971)											
	2.6	2.8	3.0	2.85	3.05	3.25	3.10	3.30	3.50	3.35	3.55	3.75
	Alyeska throughput											
8%	0.804	0.843	0.882	0.852	0.891	0.930	0.901	0.940	0.979	0.950	0.988	1.027
10%	0.943	0.993	1.042	1.005	1.055	1.104	1.067	1.117	1.170	1.129	1.179	1.228
12%	1.108	1.170	1.232	1.186	1.248	1.310	1.263	1.326	1.388	1.341	1.403	1.465
	Accelerated throughput											
8%	0.720	0.753	0.785	0.761	0.793	0.825	0.801	0.833	0.866	0.841	0.874	0.906
10%	0.822	0.862	0.903	0.872	0.913	0.953	0.923	0.963	1.003	0.973	1.013	1.053
12%	0.938	0.987	1.036	0.999	1.050	1.097	1.061	1.116	1.159	1.122	1.171	1.220
	Accelerated throughput with two-year delay											
8%	0.790	0.823	0.867	0.837	0.875	0.913	0.885	0.922	0.960	0.932	0.969	1.007
10%	0.932	0.981	1.029	0.993	1.041	1.090	1.053	1.102	1.151	1.114	1.163	1.211
12%	1.100	1.162	1.223	1.177	1.240	1.300	1.254	1.316	1.377	1.331	1.393	1.454

A Benefit-Cost Analysis

Foreign Oil as an Alternative Source of Supply

By making the same assumptions as those in chapter 1 and excluding any differences in the cost of domestic crudes between the West Coast and the states east of the Rockies, TCP benefit-cost ratios can be calculated (exclusive of opportunity cost differences) by using the same procedure employed for the TAP alternative. Algebraically, this procedure was stated in equation (4) in chapter 1.

The net benefits of a TCP can also be determined by a method analagous to that used for TAP. Several differences should be noted, however. First, the principal market will not be market i, Los Angeles. Therefore, j can be used to designate the Chicago market. Second, the external costs and benefits may not be identical for an all-land transportation system. Indeed, the analysis in chapter 2 shows that for most environmental impacts an all-land system is superior. Row vector $\bar{Y}$ may be used in place of $\bar{X}$ to reflect this and other external cost and benefit distinctions. Finally, the conclusion reached in chapter 2 was that $(B^i_{NS} = B^i_A)$; that is, the benefits to consumers were identical in market i regardless of the source of supply. This may not be true in the domestic markets east of the Rockies since they have very different institutionally imposed restrictions from those on the West Coast market. With such changes in mind, I define the net benefits of a Trans-Canadian pipeline to Chicago (market j) as follows:

$$b = (B^j_{NS} - C^j_{NS}) - (B^j_A - C^j_A) - \bar{Y}. \tag{1}$$

If similar assumptions to those in chapter 2 are used initially, then

$$B^j_{NS} = B^j_A,$$

$$\bar{Y}_\iota = 0,$$

and equation (1) reduces to

$$b^* = C^j_A - C^j_{NS}. \tag{2}$$

It is not enough to compare the cost of transporting North Slope oil to different markets in determining the most economically efficient route. The reason for this is that there may be different costs of supplying different potential markets from alternative sources of supply. Recall that net benefits for TAP to market i were defined earlier as

$$b' = C^i_A - C^i_{NS}.$$

The importance of this conclusion can be appreciated by noting that the best route is the one with greatest net benefits. Therefore, if

$$(C_A^j - C_A^i) \geq (C_{NS}^j - C_{NS}^i),$$

then

$$(B/C)_j \geq (B/C)_i.$$

That is, the optimal pipeline route is the one with the greatest resource savings compared with its next best source of supply. If both markets have the same alternative sources of supply, market j should be supplied by NS and market i by A, since this would minimize the resource cost of supplying both markets. The importance of this conclusion should be noted for the problem under consideration. In circumstances with two mutually exclusive pipeline alternatives, taking into account opportunity costs forgone means that selecting the system that yields the lower net benefits of the two alternatives results in real economic losses.

If it is assumed that in the absence of North Slope oil the long-run source of supply for the Midwest or East Coast is the Persian Gulf (as a reference crude), then the same costs per barrel that were used for the West Coast[8] can be used to calculate the benefits of North Slope oil delivered to the East Coast ($2.50 per barrel for the more conservative cost estimates). Comparable cost estimates per barrel for foreign overseas crude delivered to Chicago are about 24¢ to 25¢ per barrel greater than for either New York or Los Angeles because of Chicago's interior location. If this difference is less than any extra costs of supplying Chicago from the North Slope relative to Los Angeles, then TAP is not the most economically efficient choice.

If a Trans-Canadian pipeline is extended to New York or if exchanges between domestic sources that now supply Chicago are made for North Slope crude in the Midwest, the additional transportation costs will also be about 24¢ to 25¢ higher per barrel. Therefore, if a long-run view is taken and only foreign overseas crude is displaced, a Midwest market for North Slope oil would yield about 49¢ (24.5¢ + 24.5¢ per barrel) more net social benefits per barrel to the nation than North Slope oil brought to the East Coast. Simply stated, it would be in the nation's interest not to have Alaskan oil moving to a point farther east (at higher costs) than the most interior point supplied by imported overseas crude oil. In appendix A this approach is described diagrammatically for alternative U.S. markets.

[8] Transportation costs to either U.S. coast are about equal.

It is important to note that while net social benefits depend on whether a Midwest or East Coast market is selected, the wellhead price, and therefore the present value of net profits and taxes that are directly based on it, do not.[9] The reason for this is that the present price difference between the New York and Chicago markets is about 25¢ per barrel, which would just offset the additional transportation costs necessary to reach the more distant East Coast market.

Table 21 shows wellhead prices and benefit-cost ratios for Chicago and New York. The benefit-cost ratios were determined by dividing the more conservative costs per barrel for foreign crude used in chapter 1 ($2.49 for New York and $2.74 for Chicago) by the appropriate cost per barrel for each capital cost and production schedule case. All calculations are made at a 10 percent discount rate, except those given for 8 percent and 12 percent in a footnote.

Such computations are meaningful only when they are compared with similar estimates for TAP, which were presented in table 11 and are summarized in table 22 with the results in table 21. Each of the five variables in the table is derived when it is assumed that (1) a foreign source of supply is used as an alternative to North Slope crude in the Los Angeles (for TAP) and Chicago (for TCP-I) markets, (2) the social rate of discount is 10 percent, and (3) all the oil is consumed in the principal market selected. TAP is clearly inferior to TCP-I from the standpoint of wellhead prices and, therefore, of state taxes and corporate profits. If identical production schedules are compared, the system cost differences are relatively small. Since the costs of transporting foreign crude inland to the Midwest exceed these differences, the Midwest market represented by Chicago is an economically more efficient alternative than a coastal market for Alaskan crude, which can be supplied by foreign overseas imports more cheaply. Given the either/or situation, the benefit-cost ratios shown in table 22 indicate that TCP-I is superior to TAP. Under these circumstances, the selection of TAP over TCP-I would generate negative net social benefits, since TAP is an economically inferior alternative. If the environmental advantages of TCP-I are also considered, this conclusion is strengthened. Before making such conclusions, however, several additional benefit-cost adjustments are necessary, and the ap-

[9] Taxes per barrel are about 20 percent of this wellhead price and profits per barrel are determined by subtracting these taxes and production costs of about 25¢ per barrel. The present value of taxes and profits can then be determined by multiplying these per barrel estimates by the present value of the corresponding throughput schedule.

TABLE 21. Wellhead Price and Benefit-Cost Ratios Exclusive of Opportunity Costs at a Discount Rate of 10 Percent for a Mackenzie Valley Crude Oil Pipeline to Edmonton and Then to Chicago (TCP-I)

	Capital cost to Edmonton ($ billion, 1971)											
	2.00			2.25			2.50			2.75		
	Wellhead price per barrel[a] ($)	Benefit-cost ratio (conservative)		Wellhead price per barrel ($)	Benefit-cost ratio (conservative)		Wellhead price per barrel ($)	Benefit-cost ratio (conservative)		Wellhead price per barrel ($)	Benefit-cost ratio (conservative)	
Throughput schedule		Chicago	New York		Chicago	New York		Chicago	New York		Chicago	New York
Capital cost to Chicago—0.6 ($ billion, 1971)												
Alyeska	2.87	2.64[b]	1.90	2.80	2.48	1.81	2.74	2.33	1.72	2.68	2.21	1.64
Accelerated	2.99	3.03	2.12	2.94	2.85	2.02	2.89	2.70	1.94	2.84	2.56	1.86
Accelerated after a 2-year delay	2.88	2.67	1.92	2.82	2.51	1.83	2.76	2.36	1.74	2.70	2.23	1.66
Capital cost to Chicago—0.8 ($ billion, 1971)												
Alyeska	2.82	2.51	1.82	2.76	2.36	1.74	2.69	2.23	1.66	2.63	2.11	1.59
Accelerated	2.95	2.89	2.04	2.90	2.73	1.95	1.85	2.59	1.87	2.80	2.46	1.80
Accelerated after a 2-year delay	2.83	2.54	1.84	2.77	2.40	1.76	2.71	2.26	1.68	2.65	2.14	1.60
Capital cost to Chicago—1.0 ($ billion, 1971)												
Alyeska	2.77	2.39	1.75	2.71	2.25	1.67	2.64	2.14	1.60	2.58	2.03	1.53
Accelerated	2.91	2.76	1.97	2.86	2.61	1.89	2.81	2.48	1.81	2.76	2.36	1.74
Accelerated after a 2-year delay	2.78	2.42	1.97	2.72	2.28	1.69	2.66	2.16	1.62	2.60	2.05	1.55

[a] The corresponding calculations for the Alyeska throughput schedule (capital cost to Chicago, $0.6 billion) when the discount rate is 8% and 12% are shown below for comparison:

8%	3.01	3.10	2.16	2.96	2.92	2.06	2.91	2.76	1.97	2.86	2.62	1.89
12%	2.70	2.25	1.67	2.62	2.10	1.58	2.55	1.97	1.50	2.47	1.86	1.42

[b] The comparable estimate of the benefit-cost ratio for this case when the lower cost per barrel for Persian Gulf oil is used would be 1.91 compared with 2.64. The difference between these estimates would be approximately the same for all the other cases in this table.

TABLE 22. Summary Economic Comparison of TAP to Los Angeles and TCP-I to Chicago, Using a 10 Percent Discount Rate

	Total system costs (*$/barrel*)		Wellhead price (*$/barrel*)		Present value of taxes[a] (*$ billion, 1971*)		Present value of profits before corporate income tax[b] (*$ billion, 1971*)		Conservative case benefit-cost ratio	
Throughput schedule	Low est.	High est.	Low est.	High est.	Low est.	High est.	Low est.	High est.	Low est.	High est.
TAP										
Alyeska	0.91	1.10	2.07	2.26	1.7	1.8	5.7	6.3	2.05	2.46
Accelerated	0.83	0.98	2.19	2.34	2.2	2.3	7.5	8.1	2.29	2.71
TCP-I										
Alyeska	0.94	1.23	2.58	2.87	2.1	2.3	7.4	8.3	2.03	2.64
Accelerated	0.82	1.05	2.76	2.99	2.8	3.0	9.8	10.7	2.36	3.03
Accelerated after a 2-year delay	0.93	1.21	2.60	2.88	2.1	2.4	7.6	8.4	2.05	2.67

[a] Calculated by multiplying the wellhead price by 20% and the resultant per barrel tax by the appropriate present value of throughput.

[b] Determined by subtracting royalty, severance taxes, and production costs from the wellhead price and then multiplying by the appropriate present value of throughput.

parent irrationality of the decision made by the oil companies must be examined further.

Domestic versus Foreign Oil as a Source of Supply

In chapter 1 I assumed that the most likely alternative source of supply for North Slope crude oil in District V (the U.S. West Coast), was foreign oil. In the states east of the Rockies it may be less likely that this will be the case—at least in the short run—for several reasons. First, domestic production in these regions is more than six times the total consumption on the West Coast. Second, some production in Districts I to IV is restricted by state demand prorationing laws that hold back production from some of the more efficient wells and fields in order to maximize oil production over time and maintain a price sufficiently high to encourage and reward exploratory drilling. Third, a larger proportion of domestic production on the West Coast is based on secondary recovery techniques. Therefore, the probability of increased production from secondary recovery is relatively greater in the eastern regions of the United States. Fourth, both the West Coast and East Coast have the prospect of additional offshore discoveries and expansion of domestic sources. Fifth, the present import restrictions on the West Coast protect domestic producers at a historical price and permit foreign oil to be imported to meet the difference between the quantity of domestic plus Canadian oil supplied and the domestic quantity demanded at this price. On the other hand, imports in Districts I to IV are based on a percentage of domestic production. Accordingly, if present import quota regulations are maintained, the probability of cheaper foreign oil increasing in the West as a percentage of domestic supply is far greater than in the East if North Slope oil is not produced. That is, present institutional structures mean domestic producers east of the Rockies are more likely to expand production in a growing market.

The relative quantities produced in each section of the country are particularly important, since production east of the Rockies is more than six times greater than West Coast production. Therefore, small percentage increases in supply resulting from various incentives, such as increasing prices, new technology, and more favorable tax treatment will cause a far greater absolute increase in oil supplied for markets east of the Rockies.

Two conclusions can be drawn from the discussion above. First, foreign oil tends to be far less costly in both real and monetary terms than the marginal barrel of oil produced in the eastern part of the United States. Therefore, if the alternative to North Slope oil is more likely to be a domestic source (P_d) than a foreign source (P_f), the average cost of the alternative to North Slope oil will be greater in both monetary and real terms in the Midwest and on the East Coast than on the West Coast. Second, the market is far greater in terms of its actual size and its growth in demand; therefore, both markets will probably find their dependence on foreign oil increasing over time. From a national security standpoint, the full throughput of either crude oil pipeline will displace foreign crude.

The difference between alternatives is that because of institutional rigidities in the eastern part of the United States, North Slope crude is more likely to back out higher-cost domestic crude as well. For this reason the assumption that relatively low-cost foreign sources of supply will be used as a substitute for North Slope crude oil may not be as appropriate for Chicago as for Los Angeles. One way to account for this distinction is to estimate the probability of using domestic or foreign sources of supply in each market. The average costs of domestic and foreign sources of supply can then be combined to form a weighted average alternative cost, as follows:

$$\bar{C}^j_{DF} = P^j_d C^j_D + P^j_f C^j_F \qquad (3)$$

where

$\bar{C}^j_{DF}$ = weighted average cost of alternative source of supply in market j from domestic (D) and foreign (F) sources of supply,

$P^j_d = 1 - P^j_f$ = the probability of domestic sources being used as a substitute in market j,

P^j_f = the probability of foreign sources being used as a substitute in market j,

C^j_D = average cost of domestic crude in market j,

C^j_F = average cost of foreign crude in market j.

An additional adjustment can be made by relaxing the assumption of equal benefits to consumers of crude oil from different sources of supply. The viability of domestic and foreign production restrictions east of the Rockies will certainly be affected differently if North Slope crude rather than foreign crude is used as a source of supplying Midwest demand in the future. Another measure of benefits is that oil consumers in different markets value products made from crude oil differently. For example, gasoline is a major product of refineries on the West Coast. Gasoline is light, and the lighter the crudes, the easier (that is, least costly) it is to derive gasoline from crude oil. In the Midwest a greater mix of products is demanded by consumers. One of the main reasons for this demand is that home-heating fuels (which are relatively heavy products) are highly valued in the Midwest but not on the West Coast. Since North Slope crude is a relatively heavy crude, its value to refiners and consumers is greater in the Midwest due to its potential for greater product mix than on the West Coast. Since lighter crudes can be substituted more readily on the West Coast and since the entire world market is potentially available for import, this relative advantage of a Midwest final market for crude is important. One method of measuring this relative importance is to note that most crudes east of the Rockies are adjusted 2¢ for every degree of API specific gravity difference, while on the West Coast the penalty for heavy crudes is about 6¢ or 7¢, or about three times greater. The importance of these exceptions will be analyzed to determine their effect on benefit-cost ratios, state taxes, and the profitability of each Arctic transportation system.

The first step is to estimate the probabilities of using foreign or domestic sources of supply in each market. The greater the domestic production is in a region of the United States, the less likely is the probability of using foreign oil as an alternative source of supply. For simplicity, I assume that these probabilities (P_d^j and P_f^j) may be assigned on the basis of the percentage the full North Slope throughput would be of the approximate present domestic production plus future North Slope production. Therefore, for Districts I to IV, I will use the following weights:[10]

$$P_f^j = \frac{2 \text{ MMb/d}}{(10 \text{ MMb/d} + 2 \text{ MMb/d})} = \frac{1}{6},$$

$$P_d^j = 1 - P_f = \frac{5}{6}.$$

A similar set of weights (P_d^i and P_f^i) can be determined for the West Coast. Using the same formula as that for east of the Rockies, these probabilities will be approximately equal, since the expected TAP throughput approximately equals present "domestic" production in District V:

[10] These weights slightly exceed the present regulatory limit of one-eighth of domestic crude production utilized to determine the amount of foreign crude imports in Districts I to IV. The one-sixth therefore biases the benefit estimate downward slightly, given present institutional constraints.

$$P_f^i = \frac{2 \text{ MMb/d}}{(2 \text{ MMb/d} + 2 \text{ MMb/d})} = \frac{1}{2},$$

$$P_d^i = 1 - P_f^i = \frac{1}{2}.$$

There are insufficient data available to calculate relative domestic costs. One simple method to approximate relative cost differences is to use relative price difference as a proxy. Prices of domestic crude oil are available at the source of production. Data for making specific adjustments to account for qualitative differences in domestic crudes, such as specific gravity and sulfur content, are also available. The wellhead prices of these domestic crudes are net of transportation costs to specific markets. Since such transportation costs are real economic costs, it is appropriate to calculate the cost of these reference crudes in specific consuming markets, such as Los Angeles, Chicago, and New York. Prices in the Chicago and New York markets are determined by adding the transportation costs of crude oil from the eastern Texas-southern Louisiana Gulf region to the price of crude oil in this region. Oil from other producing areas that sell crude in this gulf region is priced at a lower amount at the wellhead to reflect the cost of transporting the crude to the gulf.[11]

Relative price differences between the markets on the West Coast and east of the Rockies will also reflect the relative cost of refining crudes for different product mixes. Since these differences are also important to consider in assessing real economic benefits, this is an additional reason to use relative price difference. Finally, domestic market restrictions for crude oil make it likely that domestic producers—in the absence of a large low-cost domestic source of supply—will produce crude oil at a high marginal cost. If prices are expected to increase if North Slope oil is not supplied to a particular U.S. market, then higher marginal cost oil will be produced than is now the case. Therefore, it is appropriate to use the present price as a measure of the average cost of additional domestic sources of supply and also to assume in this range that the average domestic cost equals the present marginal cost.

[11] For a discussion of the Gulf Coast basis of pricing east of the Rockies, see U.S. Department of the Interior, *Analysis of Economic and Security Aspects*, vol. 1, p. C-4. For an interpretation of this basis as it relates to the present controversy, see Charles J. Cicchetti and John V. Krutilla, "Comments on the Final Environmental Impact Statement for the Proposed Trans-Alaska Pipeline and the Complementary Analysis of the Economic and Security Aspects of the Trans-Alaska Pipeline" (submitted to the Department of the Interior, May 5, 1972).

If the weights and present prices are used and it is assumed that a system of "swaps" is developed so that in effect half the crude from the North Slope displaces higher-cost domestic and foreign crudes in Chicago and New York, a weighted average cost of the alternative similar to that shown in equation (3) can be determined as follows:

	Midwest	East Coast
	(dollars per barrel)	
Present marginal domestic cost (assume marginal = average)	3.81	4.06
Persian Gulf (conservative)	2.75	2.50
Weighted average (5/6 and 1/6)	3.63	3.80
Average cost of alternative source of supply east of the Rockies	3.715	

When a mixed domestic and foreign alternative source of supply is used, a similar adjustment for the Los Angeles market can be calculated as follows:

$$1/2\ (\$3.17) + 1/2\ (\$2.50) = \$2.835.$$

These estimates of $3.715 and $2.835 per barrel can be compared with the per barrel benefit estimates of $2.74 and $2.49, respectively, for the markets east of the Rockies and on the West Coast. The latter values were based on a long-run view that foreign crude was the only alternative source of supply.

Additional Benefit-Cost Adjustments

In the computations that follow, several additional parameters are reviewed. First, as mentioned earlier, North Slope oil is medium in sulfur content. The domestic crudes that supply the Midwest market, however, are generally lower in sulfur content than domestic crudes on the West Coast.[12] To adjust for this difference, crude prices might be reduced east of the Rockies. A difference of 11¢ per barrel exists between the posted wellhead of West Texas Sour and West Texas Intermediate Sweet.[13] A second possible adjustment is that North Slope oil

[12] Letter to the author from E. L. Patton, President of Alyeska Service Company, September 30, 1971.

[13] *Platt's Oilgram Price Service, Crude Oil Supplement,* vol. 49, no. 164–B (August 25, 1971).

may be lighter than 26.9° API, in which case the penalty charged against it may be too severe for the West Coast relative to the Midwest. This adjustment may reduce relative prices by an additional 5¢–10¢ per barrel. To account for both factors, the various benefit estimates will be reduced by 18¢ per barrel. This combined adjustment is identified as the Alyeska price adjustment factor in table 23.

The adjustment for sulfur content and specific gravity penalties reduces the reference price in the Midwest to $3.63 per barrel. It is important to note, however, that this adjustment may be in the wrong direction, since early analysis of North Slope crude indicated it scored particularly high with a hydrocracking refining technique, which is now being initiated in the Midwest.[14] On the basis of that finding, a reference price of $3.93 per barrel would be in order. As a final point on prices, the Department of the Interior used a maximum price of $3.50 per barrel for reference purposes in the Midwest.[15] This has recently been increased to $3.60 per barrel,[16] and the East Coast price has always been consistent with the higher prices used in the analysis above. Therefore, the case labeled Alyeska Adjustment in table 23 can also be interpreted as the new Interior Department maximum Midwest price case,[17] and the Midwest price used in the analysis of $3.81 per barrel without adjustment is high or low depending on future conditions, but it certainly falls in the most likely range of $3.63 to $3.93 per barrel for the Chicago market.

The environmental advantages of the Alaskan Highway route (TCP-IV) were presented earlier. Since this route is longer than the Mackenzie Valley route (TCP-I), Walter Parker estimated it would cost about $400 million more in 1971 dollars.[18] When I take this difference into account below, I also increase the operating costs of this longer route proportionately.

[14] R. R. Roselius and J. H. Steffens, "North Slope Oils Score High with Hydroprocessing," *Oil and Gas Journal*, May 17, 1971.

[15] Nicolai Timenes, Jr., "Analysis of Transportation Alternatives," app. C, *Analysis of Economic and Security Aspects*, vol. 1, *Summary*.

[16] William Vogeley, "Comments on Trans-Alaska Pipeline and Alternatives Submitted by the Environmental Defense Fund to the Department of the Interior and Cicchetti-Krutilla" (memo to the undersecretary submitted to the Department of the Interior, May 8, 1972).

[17] For a complete discussion of the prices used for reference purposes in each market by various analysts, see Cicchetti and Krutilla, "Comments on the Final Environmental Impact Statement."

[18] Walter B. Parker, *A Comparison of Prudhoe Oil Costs via Valdez or via a Mid-Canada Pipeline*, Federal Field Committee for Development Planning in Alaska (Anchorage, March 1971).

TABLE 23. Benefit-Cost Ratios (Conservative Prices) for Trans-Canadian Alternatives to Chicago and the Trans-Alaska Pipeline to Los Angeles at a 10 Percent Discount Rate

	Capital cost to Chicago ($ billion, 1971)								Capital cost to Valdez ($ billion, 1971)			
	2.8	3.55	2.8	3.55	3.2	3.95	3.2	3.95	2.0	2.5	2.0	2.5
	Mackenzie Valley Rte. (TCP-I)				Alaska Highway Rte. (TCP-IV)				Trans-Alaska Rte. (TAP)			
Appreciation in estimated cost	Accelerated throughput		Accelerated after 2-yr. delay		Accelerated throughput		Accelerated after 2-yr. delay		Accelerated throughput		Alyeska throughput	
Foreign alternative source of supply												
0%	2.89	2.46	2.54	2.14	2.59	2.24	2.27	1.95	2.55	2.29	2.31	2.05
10%	2.62	2.23	2.31	1.95	2.36	2.04	2.07	1.77				
20%	2.41	2.05	2.12	1.78	2.16	1.87	1.89	1.62				
Domestic and foreign source of supply[a]												
0%	4.18	3.56	3.68	3.11	3.75	3.24	3.29	2.82	3.22	2.91	2.94	2.61
10%	3.83	3.26	3.37	2.85	3.44	2.98	3.02	2.59				
20%	3.54	3.01	3.12	2.62	3.18	2.75	2.78	2.39				
Domestic and foreign with Alyeska price adjustment factor												
0%	3.94	3.35	3.46	2.92	3.53	3.05	3.11	2.66				
10%	3.61	3.08	3.18	2.68	3.24	2.80	2.84	2.44				
20%	3.34	2.84	3.94	2.48	3.00	2.59	2.62	2.25				

[a] Similar benefit-cost ratios for New York can also be calculated. For a case without cost appreciation, the following ratios would result:

3.55	3.02	3.11	2.62	3.18	2.74	2.78	2.38

Note that the 3.55 for New York is based on the same assumptions as those for the 4.18 for Chicago. If the benefit-cost ratios for an Alyeska schedule are important, they are approximately equal to the ratios shown for an accelerated schedule after a two-year delay.

The last factor that will be considered here tests the sensitivity of the results of an assumption that the real costs of the Canadian alternatives have been underestimated. This could be due to uncertain and inaccurate cost data or possibly to new taxes or other real costs imposed by Canada, such as a guaranteed capacity for Canadian oil. Such decisions might result in U.S. consumers of oil paying higher costs or taxes to the Canadian government and thus would represent real costs. All adjustments of this kind would be in the same direction; that is, the cost would increase. Therefore, the sensitivity of the results to appreciation in all costs by 10 percent and 20 percent will be examined.[19]

In table 23 the benefit-cost ratios are shown adjusted for the factors above, and similar calculations for TAP (summarized from table 11) are given. Only at the highest capital cost per barrel, as in the TCP-IV case (with capital costs of $3.95 billion) with a two-year delay and 20 percent cost appreciation (making capital cost equal to about $5 billion) do the benefit-cost ratios shown for a mixed domestic and foreign alternative fall to levels as low as the results for TAP at the lower end of the spectrum of capital cost and the Alyeska production schedule. If costs have not been understated, the benefit-cost ratios for TCP alternatives always are greater than or equal to the TAP alternative in all but one case. That one exception occurs when the lowest capital cost estimate for TAP is compared with the highest capital cost estimate for TCP-IV.

Table 24 presents an index that is easier to use for comparison purposes. It is the net benefit estimate based on average capital costs for the three main routes, TCP-I, TCP-IV, and TAP. The table shows that in all cases, including TCP-IV, with a two-year delay and a 20 percent increase in costs the net benefits exceed—in some cases by a factor of about two-thirds—similar estimates for TAP. A second significant difference revealed by this table is that net benefits for TCP-I are greater than those for TCP-IV by about $0.5 billion in present value of 1971 dollars. The difference is the amount that any environmental advantages attributed to TCP-IV in chapter 2 would have to equal or exceed (in the sense of a discounted time stream of environmental cost differences) in order for TCP-IV to be preferred over TCP-I.

While tables 22 and 23 show that the social value or net benefits of a TCP alternative exceed comparable TAP estimates, it is also useful

[19] These adjustments would amount to higher costs or taxes paid to the Canadian government of approximately 10¢ to 25¢ per barrel.

TABLE 24. Net Benefit Estimates for the Three Principal Pipeline Alternatives, Using a 10 Percent Discount Rate and Average Capital Costs

	Mackenzie Valley (TCP-I)		Alaska Highway (TCP-IV)		Trans-Alaska (TAP)	
	Accelerated	Accelerated after 2-yr. delay	Accelerated	Accelerated after 2-yr. delay	Accelerated	Alyeska
Case I						
Av. capital cost (*$ billion, 1971*)	3.2		3.6		2.25	
Per barrel net social benefits (*$/bbl.*)	2.53	2.40	2.43	2.29	1.66	1.55
Present value of net social benefits (*$ billion, 1971*)	12.6	9.9	12.1	9.5	8.3	6.3
Undiscounted net social benefits, assuming 15 billion barrels are produced (*$ billion, 1971*)	38.0	36.0	36.5	34.2	24.9	23.3
Case II—Operating and Capital Costs of Trans-Canadian alternatives increase by 20%						
Av. capital costs (*$ billion, 1971*)	3.9		4.3		2.25	
Per barrel net social benefits (*$/bbl.*)	2.34	2.18	2.23	2.05	1.66	1.55
Present value of net social benefits (*$ billion, 1971*)	11.6	9.0	11.1	8.5	8.3	6.3
Undiscounted net social benefits, assuming 15 billion barrels are produced (*$ billion, 1971*)	35.1	32.7	33.5	30.8	24.9	23.3

to ascertain what effect variations in the parameters that were adjusted above would have on the tax revenue of Alaska and the net profits of the oil producers. Also, in appendix C, George Rogers describes the likely impact of each alternative route on employment in Alaska. In table 25 these results are shown in present value form along with directly comparable TAP results derived from table 11.

It is important to note that state revenue and oil company profits generally will not be affected by the adjustments to real costs and benefits made in tables 23 and 24 when domestic alternative sources of supply are included in the analysis. Therefore, whether or not "real" benefits are adjusted to reflect these differences will not affect the financial desirability of each alternative.[20]

An important conclusion to be drawn from table 25 is that tax revenues will always be greater for Alaska under a TCP when TCP production schedule estimates are compared with similar TAP estimates.[21] This would be so even if the highest capital cost alternative were to be built and costs were to increase by as much as 20 percent per barrel. Under more favorable cost conditions, the state might find its present value of expected tax receipts to be substantially higher under a TCP alternative, which would mean approximately $1 billion more in the present value of taxes (undiscounted this could amount to a difference of nearly $5 billion over the life of the North Slope field).

The second conclusion to be drawn from table 25 is that the oil producers would always receive more in present value of net profits for the TCP alternative, given similar production schedules. Moreover, it is only when the accelerated schedule, lower-cost TAP alternatives are compared with the two-year delayed, highest-cost TCP-IV route and a 20 percent appreciation in cost that the firms' present value of profits approaches a point of indifference between TCP-IV and TAP. In more favorable cases for the TCP alternatives, the oil producers could expect to receive between $3 billion and $5 billion more in the present value of additional profits before corporate income taxes (approximately $16 billion undiscounted) than they would for comparable TAP capital cost and production schedule cases.

[20] The adjustment of 18¢ per barrel for possible lighter North Slope crudes and sulfur content differences would have a small percentage effect on the wellhead prices and therefore on the taxes and profits. But since this adjustment is both uncertain and perhaps an overestimate of any adjustment of this kind, it will not be used.

[21] "Similar" means accelerated is compared with accelerated and Alyeska is compared with accelerated with two-year delay.

TABLE 25. Present Value of Taxes and Net Profits at a Discount Rate of 10 Percent

(*billions of dollars*)

Appreciation in estimated costs	Accelerated throughput: Present value of taxes	Accelerated throughput: Present value of net profits	Accelerated with 2-yr. delay: Present value of taxes	Accelerated with 2-yr. delay: Present value of net profits	Accelerated throughput: Present value of taxes	Accelerated throughput: Present value of net profits	Accelerated with 2-yr. delay: Present value of taxes	Accelerated with 2-yr. delay: Present value of net profits
	Mackenzie Valley Rte. (TCP-I)							
Capital cost to Chicago ($ billion, 1971)	← 2.8 →				← 3.55 →			
0%	2.94	10.5	2.32	8.3	2.79	9.9	2.17	7.7
10%	2.85	10.1	2.24	7.9	2.69	9.5	2.08	7.3
20%	2.77	9.8	2.16	7.6	2.58	9.1	2.02	6.9
	Alaska Highway Rte. (TCP-IV)							
Capital cost to Chicago ($ billion, 1971)	← 3.2 →				← 3.95 →			
0%	2.84	10.1	2.22	8.0	2.69	9.5	2.08	7.3
10%	2.74	9.7	2.13	7.5	2.58	9.1	1.97	6.9
20%	2.65	9.3	2.04	7.2	2.47	8.6	1.87	6.5
	Trans-Alaska + Tankers (TAP)							
Capital cost to Valdez ($ billion, 1971)	← 2.0 →				← 2.50 →			
0%	2.28	7.9	1.78[a]	6.1[a]	2.18	7.5	1.68[a]	5.7[a]

[a] Alyeska schedule.

If, as the economic analysis in this chapter reveals, the TCP alternatives are very likely to be far superior from the standpoint of net benefits (benefit-cost ratios), tax revenue for Alaska, and profits for oil producers, the question must be asked: Why was the TAP alternative chosen over the TCP alternative? I will address that question in succeeding chapters by analyzing the various U.S. and world markets for petroleum in a more disaggregate fashion.

4

Some Further Economic and Institutional Comparisons of Alternative Transport Systems

In the previous chapter, I pointed out that the highest-priced U.S. markets for oil were in the Midwest and on the East Coast. It is sometimes assumed, however, that Alaskan oil reaching Edmonton would be divided between the high-priced eastern markets and the West Coast markets. In that case, additional economic and environmental costs entailed in shipping crude oil from Puget Sound to California would have to be added to the Trans-Canadian Pipeline (TCP) costs considered earlier. Since tanker traffic and port activities have not entered into the analysis of the TCP alternatives thus far, these alternatives were considered to be environmentally superior to the Trans-Alaska Pipeline (TAP) from this standpoint. Accordingly, a decision to transport some of the North Slope oil to California via Puget Sound would become an important factor in selecting an optimal route.

A comparison of the environmental impacts of different transportation systems depends on the following uncertain factors: (1) the amount of oil diverted westward, (2) whether a new oil pipeline or tankers would be used to transport oil to California, and (3) whether more oil would be transported via Puget Sound to California with a TCP alternative than into Puget Sound with the Trans-Alaska Pipeline (TAP). The last factor in turn depends on (1) the future status of Canadian overland exemptions in the Puget Sound area, (2) the possible displacement of Albertan crude oil in the Trans-Mountain line to Edmonton and the Interprovincial Pipeline to the Midwest, and (3) the possibility of a new pipeline to the Midwest across the northern part of the United States.

The Effects of an Edmonton-to-Seattle Alternative

Before proceeding further it is important to evaluate economically the alternative of splitting the North Slope oil at Edmonton. Table 26 shows the net benefits and net revenue that would be expected if 40

TABLE 26. Net Benefits (Assuming a Mixed Domestic and Foreign Alternative) and Net Revenue per Barrel for TCP and TAP at Average Capital Costs and a Discount Rate of 10 Percent

(dollars per barrel)

	Throughput schedules	
Route	Accelerated	Accelerated (2-yr. delay)
TCP-I with 40% to District V		
Net benefits	1.53	1.39
Net revenue	1.92	1.74
TCP-IV with 40% to District V		
Net benefits	1.43	1.28
Net revenue	1.76	1.62
TCP-I with 100% to the Midwest or East Coast		
Net benefits	2.53	2.40
Net revenue	2.63	2.50
TCP-IV with 100% to the Midwest or East Coast		
Net benefits	2.43	2.29
Net revenue	2.53	2.39
TCP-I weighted average (40% to District V; 60% east of the Rockies)		
Net benefits	2.13	1.99
Net revenue	2.34	2.20
TCP-IV weighted average (40% to District V; 60% east of the Rockies)		
Net benefit	2.03	1.89
Net revenue	2.22	2.08
TAP to Los Angeles		
Net benefit	1.65	1.55[a]
Net revenue	1.99	1.89[a]

NOTE:

Capital cost: TCP-I	($2 billion + $2.75 billion) divided by 2
TCP-IV	($2.4 billion + $3.15 billion) divided by 2
Edmonton to Chicago	$0.8 billion
Edmonton to Seattle	$0.7 billion
TAP	$2.25 billion

Assume that for those cases in which 40% of North Slope oil will be sent to the West Coast the cost of oil from Seattle to Los Angeles is 20¢ per barrel.

[a] These estimates are based on the Alyeska throughout schedule rather than the accelerated plan with a two-year delay. The results would be within one or two cents of these figures, however.

percent of a TCP throughput were diverted to the West Coast and 60 percent to eastern U.S. markets. Similar estimates for a totally (100 percent) eastern U.S. alternative are also shown.

A comparison of these estimates indicates that net benefits of approximately \$1 and additional taxes and profits of more than 70¢ per barrel would be realized by moving all the North Slope oil eastward from Edmonton. Given the probable deleterious environmental effects in Puget Sound, it is highly improbable that Alaska, Canada, or the oil companies would want to send oil from Edmonton to the West Coast.[1] But to quantify the differences more precisely, the estimated weighted net benefits and net revenues for the TCP when the oil is split at Edmonton, with 40 percent diverted to the West Coast and 60 percent to markets east of the Rockies, are also shown in table 26. A comparison of similar throughput schedules reveals that dividing Alaskan oil at Edmonton and sending 40 percent to the West Coast and 60 percent to the East would mean that net benefits would be 40¢ to 50¢ per barrel greater and that net revenues and taxes would be 20¢ to 35¢ per barrel greater for TCP alternatives relative to TAP. Accordingly, unless only a small amount of Alaskan oil were transported to Puget Sound via TAP and as much as 800,000 barrels per day[2] were transported via Puget Sound to California under any of the TCP alternatives—which would result in an environmental cost of 40¢ to 50¢ per barrel—the TCP alternatives would be economically and financially superior to TAP, even if the uneconomic decision to move Alaskan oil westward were made.

The Canadian Share of the Daily Throughput of a Trans-Canadian Pipeline

One objection to a TCP route is that the Canadian government may insist that a certain portion of the capacity of a Trans-Canadian Pipeline must be set aside for Canadian Arctic oil. While this objection may seem reasonable, several important qualifying factors must be considered. First, the present North Slope proven reserves (approximately 10 billion barrels) are below the undiscounted cumulated

[1] One possible exception is Atlantic Richfield (ARCO), which has constructed a new refinery at Cherry Point, Washington. It is unlikely, however, that ARCO would want any of its North Slope oil in excess of the oil required (approximately 100,000 barrels per day) for this new refinery to flow westward. Furthermore, optimal "swaps" could probably be made with other West Coast oil companies to meet the needs of the Cherry Point refinery.

[2] Forty percent of 2 million barrels per day.

throughput (approximately 17.5 billion barrels) used in the analysis in the preceding chapters. Therefore, sharing the capacity of an Arctic pipeline with Canada may be necessary to meet the production schedules utilized in the previous analyses. Since average costs and net benefits per barrel depend on the expected accumulated throughput, the potential availability of "new" oil discoveries on or near the proposed pipeline alternatives must be considered.

Prospects for oil in the Canadian Arctic are excellent, but explorations and discoveries there still lag behind those on the North Slope. If the most likely site of new discoveries is in the Canadian Arctic rather than in Alaska, additional pipeline construction from new Canadian oil fields would be necessary in order to connect into the TAP transportation system. On the other hand, TCP–I would pass directly through the most likely areas for Canadian oil discoveries. Furthermore, in the TCP–IV case, additional costs were added to account for the construction of a connecting pipeline that would follow the route of the new Demster Highway.

A second factor that would affect any negotiations concerning the proportions of Canadian-U.S. pipeline capacity is that Canadian oil and gas development depends on U.S. assurances of an open market for crude oil in the states east of the Rockies. Raising capital would be most difficult for Canada without such assurances. Recent governmental decisions in both countries highlight these points of international economics. For example, the United States has withdrawn the Canadian crude oil overland exemption and has fixed a limit on the amount of Canadian crude oil that can enter Petroleum Administration for Defense (PAD) Districts I to IV,[3] and the Canadian National Energy Board has decided that Canadian natural gas in the lower provinces must be conserved until Arctic gas and appropriate transportation systems become more certain.[4] A final factor that would affect negotiations between Canada and the United States is the possible Canadian loss of U.S. West Coast markets if TAP were to be constructed and Alaskan oil were to displace Canadian oil in the Puget Sound market area.

Concern over the assignment of definite proportions of the capacity of a Trans-Canadian pipeline is therefore a highly oversimplified proxy for more serious possible points of conflict. It is important to recognize the need for Canadian-U.S. governmental negotiation to determine a

[3] "Nixon Slows Oil Imports from Canada," *Oil and Gas Journal*, March 16, 1971.

[4] "Canada Rejects Increase in Gas Exports," *Oil and Gas Journal*, November 29, 1971.

pipeline route that would be beneficial to the economies and environments of both countries. In the absence of such negotiations the oil companies have tended to assume the worst, such as delays of two years, additional taxes, and pipeline throughput reductions.

The three major leaseholders on the North Slope may not be in full agreement on these issues. For example, Humble Oil is a subsidiary of Standard Oil of New Jersey and a major leaseholder on the North Slope. At the same time, a sister subsidiary, Imperial Limited of Canada, is actively involved in Canadian Arctic exploration. Humble would therefore tend to be indifferent about whether Canadian or Alaskan oil made up its throughput share.[5] Taxes paid to Canadians, however, would represent a real cost to the United States, and Canadian crude would therefore be less beneficial to the United States than Alaskan crude with similar properties. But this differential is likely to be less than the cost additions of 10 percent and 20 percent per barrel that were added to the cost estimates of the TCP alternatives in chapter 3.

Another major leaseholder on the North Slope is British Petroleum and its subsidiaries, BP Oil Corporation and Standard Oil Company (Ohio), or SOHIO. If it were possible to net out the profits received by non-U.S. citizens who are owners of the parent company, such extra-national payments would be real costs to the U.S. economy and should also be netted out of both the TAP and TCP alternatives. While such an adjustment would be consistent with treatment in the analysis above of taxes paid to foreign countries, it must be emphasized that it would be correct to subtract the costs of foreign trade only if the benefits of foreign trade were also considered. But if the consistency-of-treatment argument prevails, royalty and income taxes paid by Humble's Canadian subsidiary should be subtracted from the net benefit estimates that were used in chapter 3 for TCP, and the profits of more than $1 per barrel—when paid to non-U.S. citizens—should be removed from the TAP alternative. On net, therefore, these adjustments favor the TCP alternatives rather than TAP.

Moreover, unlike the other major leaseholders, British Petroleum has an added incentive to build the TAP alternative. In January 1970 its U.S. subsidiary, BP Oil Corporation, merged with SOHIO, gaining over 25 percent of SOHIO's stock. This percentage will increase to 54 per-

[5] Actually, the royalty and severance taxes are less in Canada, approximately 10 percent. When half the field has been produced, however, this rate is subject to possible additional upward adjustment. Therefore, Canadian crude may be preferred as a way of avoiding taxes in the early period of production but may cost more per barrel in future years.

cent when BP's North Slope production reaches 600,000 barrels per day.[6] If, under extreme conditions, Canadian oil were to make up 50 percent of the total throughput of 2 million barrels per day of a TCP alternative and BP were to produce 50 percent of the allowable pipeline capacity (1 million barrels per day) for North Slope oil, the company would fall short of its necessary 600,000 barrels per day by 100,000 barrels. While such an eventuality is unlikely, it suggests that the oil companies on the North Slope may have different objectives.

The adjustments made in chapter 3 for delays, the higher-cost Demster Highway connecting link, and the possible cost appreciation of 10 percent and 20 percent probably more than account for any additional real costs that would result from using some Canadian oil along with Alaskan oil in the TCP alternatives. Such adjustments—even when all three were made at the same time—did not change the basic conclusion that the TCP alternatives were economically and financially superior to TAP. Therefore, if the issues raised in this section were quantified, that basic conclusion concerning the superiority of a TCP system would still hold.

Equity and Consumer Surplus Considerations for U.S. Consumers of Crude Oil

In chapter 3, I restricted my consideration of the benefits of various pipeline alternatives to the average costs of domestic and foreign oil in the markets on the West Coast and east of the Rockies. The equity effects and consumer surplus benefits of each pipeline alternative were omitted. While care must be taken to avoid double counting, these additional issues are quantitatively significant and should be considered.

Table 27 shows the present prices for crude oil of comparable quality on the West and East coasts and in the Midwest. In chapter 3 I used these relative price differentials as a measure of the relative domestic average (presumed to equal marginal) cost differentials in each part of the country. Greater costs per barrel at the margin are partly due to higher transportation and production costs.

Some part of the price differences shown in the first line of table 27 is due to domestic (state prorationing) and foreign (import quota) supply restrictions. The benefits of state prorationing restrictions fall in favor of the taxpayers of producing states and the owners of oil

[6] BP Alaska, Inc., *North Slope Alaska: Man and the Wilderness* (Washington, D.C., 1971).

TABLE 27. Some Equity Effects of Route Selection

(dollars per barrel)

	West Coast	Midwest	East Coast
Prices now	3.17	3.81[a]	4.06
Prices if Alaska pipeline is built[b]	2.38	3.81[a]	4.06
Prices if a Canadian pipeline is built[b]	3.17	3.49[c]	3.73[c]

[a] Prices in the Midwest normally should be 25¢ per barrel lower than on the East Coast because of increased transportation costs to the East.

[b] Assuming that the price elasticity is unity and that half of the price decrease is passed on and, in the case of TAP, that all the oil is supplied to the West Coast, District V.

[c] Assuming that half the oil from the North Slope—1 million barrels per day—would be shipped to the East Coast.

companies. At the same time, some of the costs of such "conservation" practices are passed on in the form of higher prices to the consumers of oil who reside in states that do not produce oil.

Similarly, the national security benefits associated with restricting foreign oil affect all Americans equally. But the price differentials shown in table 27 indicate that all Americans are not paying equally in the form of similar crude oil prices for this national security benefit. The delivered price of crude oil from the Persian Gulf is approximately equal on the U.S. West and East coasts, yet there is a price differential of approximately 90¢ per barrel for oil similar in quality to North Slope crude.[7] Therefore, an additional economic basis for a comparison of alternative pipeline routes is to determine the effects of each route on the present inequities in crude oil prices in the United States.

Assume that the price elasticity of demand is equal to unity. To measure the effect of an additional 2 million barrels per day (full capacity of either TAP or TCP alternatives) on the quantity of oil supplied at the 1971 equilibrium prices and quantities, assume that the supply of crude oil is restricted in domestic markets to its present levels. Further assume that since oil companies both buy and sell oil half of any price change would be passed on to consumers.

The prices that would result in the three reference markets under these assumptions are shown in table 27. This comparison leads to the conclusion that if TAP were built instead of TCP the inequities that now exist in the form of higher relative prices paid by oil consumers in non-producing Midwest and East Coast states would be significantly increased. On the other hand, the construction of a TCP system would

[7] If lighter weight crudes, such as 30° API were used, this differential would fall to approximately 60¢ per barrel.

begin to redress these present inequities, and while prices are not expected to actually decline in either market, a reduced relative rate of price increases should favor Midwest consumers. In the future, therefore, remaining price differentials might be attributed in large part to transportation cost differences from domestic sources of supply rather than to present domestic and foreign supply constraints.

A precise measurement of the equity effects of each alternative is impossible without a more completely defined social welfare function. Since these equity effects, however, serve only to reinforce the conclusions based on the productive efficiency benefits determined in chapter 3, it is obvious that the advantages of the TCP alternatives are understated when such equity differentials are excluded, because they also favor the TCP alternatives.

Whether the present inequities between oil consumers in District V and oil consumers in the Midwest or on the East Coast would actually increase depends on several additional factors that I will discuss in the next chapter, since such price adjustments would also reduce profits. Clearly, however, these inequities would become worse, other things remaining the same, if TAP were built, and from that standpoint alone a route through Canada is preferable to TAP.

In this discussion I have first assumed institutional rigidities in the Mandatory Oil Import Quota Program east of the Rockies. Second, I have assumed for the purpose of this demonstration that domestic demand and supply do not change over time—a highly unrealistic assumption but one that allows me to keep the analysis simple. Furthermore, as I will show later, changes in this assumption tend to affect TAP adversely relative to the TCP alternatives. Finally, I implicitly assumed—perhaps naively—that alternatives other than price competition would not be used in the future if excess supply occurred, but later I will consider the equity issues raised by some of these alternatives.

Natural Gas Pipeline Considerations

Although the possible construction of a joint natural gas and crude oil pipeline has been mentioned, the implicit assumption thus far in the analysis has been that only a single-purpose, or crude oil, pipeline would be constructed. But the economic advantages of delivering Arctic oil east of the Rockies instead of to the West Coast would also be realized by a Trans-Canadian natural gas pipeline.

Table 28 shows (1) the cost of transporting Arctic natural gas to

TABLE 28. Estimated Costs of Natural Gas from the Arctic for Alternative Markets

(cents per thousand cubic feet)

	Cost
West Coast market	
Production[a]	10
Alaskan pipeline[b]	11–13
Liquefication and regasification[c]	25
Foreign tankers[d]	17
Average total cost with foreign tankers	64
Domestic tankers[e]	30
Average total cost with U.S. tankers	77
Present West Coast prices:[f]	
Southern California	36
Northern California	30–35
East Coast market	
Production[a]	10
Direct Canadian pipeline to the Midwest from Canadian gas fields	32–39
Average total cost	46
Pipeline from North Slope[g]	
To Midwest	51
To East Coast	61
Average total cost:	
To Midwest	61
To East Coast	71
Present average prices:[h]	
Chicago	33
New York	42
Alternative sources of supply	
Imported liquefied natural gas:[i]	
East Coast	60–90
Midwest	70–$1.00
Synthetic natural gas from coal:	
Expected future (1976–80)[j]	32.9–56.4
Present[k]	80

[a] See Federal Power Commission, Bureau of Natural Gas, *A Staff Report on Natural Gas Supply and Demand* (September 1969), especially p. 51.

[b] These estimates are based on an assumed present value of pipeline costs of $1 million per mile and a throughput schedule of 3 billion cubic feet per day at a 10% discount rate, with the throughput schedules proportional to the crude oil production rates used above. See: Richard A. Rice, *The North American Arctic Coast Energy Transport and Resource Distribution Problems* (Carnegie-Mellon University, Transport Research Institute, February 1971). See also *The Oil and Gas Journal,* May 17, 1971, p. 88.

[c] See FPC, *Staff Report on Natural Gas,* p. 51.

[d] Based on approximations that were made using the contracted price of 52¢ per million BTU (per MCF) for liquefied natural gas delivered from Cook Inlet to Tokyo. Ibid., p. 62.

[e] The higher costs for U.S. flagship crude oil tankers are used to approximate the cost of delivering liquefied natural gas to the U.S. West Coast.

the U.S. West Coast, (2) the cost of transporting it completely overland by pipeline to the Midwest and East Coast, and (3) the cost of several alternative sources of supply. At the present time gas prices are higher and more gas is consumed in the Midwest and on the East Coast than on the West Coast. Furthermore, higher-cost foreign gas or synthetic gas are more likely to be used in eastern markets than gas available at either the present domestic price or a future price that would include the costs of transporting Arctic gas to these markets. Under such circumstances it would be optimal to use Arctic natural gas, even if capital investments of $2.4 billion to $3 billion are required before becoming dependent on foreign or synthetic gas.

Two qualifications are in order. First, domestic sources of supply in the lower forty-eight states could probably be expanded at a lower real cost per thousand cubic feet (MCF) than either Arctic gas or foreign liquefied natural gas. At the present time domestic gas prices are regulated by the Federal Power Commission, whose restrictions cause the supply of domestic natural gas to be generally unresponsive (inelastic) to higher-cost alternatives. Second, liquefied natural gas as a relatively non-polluting fuel for transportation is gaining some proponents; therefore it may have a premium value that is not reflected by the relative prices and costs shown in table 28.

An analysis of natural gas alternatives is an important consideration in the selection of the optimal location of a crude oil pipeline in either Alaska or Canada. Expected real economic benefits from the construction of the natural gas pipeline are based on the avoidance of higher-cost supply alternatives.[8] Any cost saving from the construction of a joint oil and gas pipeline corridor should be added, along with any associated benefits generated by an Arctic natural gas pipeline, to each of

[8] There would also be a saving of several hundred million dollars in gas reinjection investment if oil and gas were produced jointly on the North Slope

---

[f] See *The Oil and Gas Journal*, February 1, 1971.

[g] This estimate is based on a presumed pipeline length of 3,400 miles, with an assumed cost of 1.5¢ per MCF per 100 miles for the last 1,000 miles and $1 million per mile for the first 2,400 miles.

[h] See FPC, *Staff Report on Natural Gas*, p. 71, in which Midwest prices are estimated in the 29¢–40¢ range and East Coast prices in the 42¢–69¢ range. These estimates have been increased by approximately 7.5¢ per MCF to reflect recent increases in the Southern Louisiana wellhead prices. See *The Oil and Gas Journal*, Oct. 4, 1971, p. 65; July 12, 1971, p. 31.

[i] See *The Oil and Gas Journal*, July 12, 1971, p. 31. For estimates that exceed $1 per MCF, see ibid., Oct. 4, 1971, p. 65, and Aug. 23, 1971, p. 58.

[j] See FPC, *Staff Report on Natural Gas*, p. 68.

[k] See *The Oil and Gas Journal*, July 12, 1971, p. 31.

the oil pipeline benefit estimates. Since cost savings over likely alternatives, domestic prices, and expected transportation costs all favor Midwest and East Coast markets over the West Coast, such adjustments would greatly increase the relative economic value of the TCP alternatives in comparison with TAP. This conclusion is especially true for TCP–I, which would pass closer to present Canadian natural gas fields than the other routes. Furthermore, cost savings are not restricted to construction and operation costs but would also include any environmental savings from constructing both pipelines at the same time and in a common transportation corridor.

The effect of such an adjustment on the benefit estimates used for the TCP alternatives is to increase the present value of crude oil by at least 10¢ to 30¢ per barrel, or about $1 billion discounted at 10 percent and $5 billion undiscounted—assuming production ratios are 1 barrel of crude to approximately 1 MCF of natural gas. This adjustment is the estimated cost saving over synthetic and liquefied natural gas alternatives in the Midwest. Any cost saving in crude oil pipeline construction and gas reinjection equipment should also be added to the net benefits per barrel of crude oil. While the environmental savings may be more difficult to quantify, several factors are important, such as a common roadbed, savings in gravel borrowings, and a single corridor for a major portion of the route. The quantifiable and unquantifiable benefits should be added to the TCP alternatives if a parallel gas line is contemplated.[9]

The important qualifications introduced in this chapter to test the sensitivity of the conclusions indicate that there are strong economic, financial, and environmental reasons for selecting a completely overland direct crude oil pipeline to the U.S. Midwest and East Coast rather than the combined TAP-tanker system. They further show that in the conclusions reached thus far the advantages of the TCP alternatives were underestimated relative to TAP.

[9] Richard Rice has discussed additional cost savings and economic benefits in a case study in which he considers a joint natural gas and crude oil pipeline constructed in the roadbed of a railroad track. See Richard A. Rice, *The North American Arctic Coast Energy Transport and Resource Distribution Problems* (Carnegie-Mellon University, Transportation Research Institute, February 1971).

5

Supply and Demand Imbalances

For the quantity of oil demanded and supplied to be in equilibrium at the present prices in each domestic market, changes in demand and supply must be offsetting. The implicit assumption thus far has been that the increased supply of North Slope oil to District V would be offset by a combination of factors, such as growth in demand, a decline in domestic production in District V, and changing institutional arrangements for foreign imports.

The results of the Trans-Alaska Pipeline (TAP) analysis are particularly sensitive to such assumptions, since District V receives separate treatment under the Mandatory Oil Import Quota Program,[1] and the growth in demand in District V has been below the national average in recent years, although depressed economic conditions on the West Coast may be the cause. Furthermore, West Coast oil consumption is much lower than that east of the Rockies. Therefore, any excess in the quantity of oil supplied is likely to have an important effect on profits, tax revenues, and economic efficiency.

In recent years the growth in demand for oil east of the Rockies (Districts I to IV) has been about 5 percent per year. If this rate continues, by 1975 the increase in demand would be 3.4 million barrels per day (MMb/d), which is far in excess of even the greatest production schedule (2 MMb/d) now contemplated for Arctic oil. Even if domestic production east of the Rockies were to grow by only 2 percent per year, by 1975 the excess quantity of oil demanded (at present prices) would be 2 MMb/d. This increased demand east of the Rock-

[1] See app. A.

ies could be met by North Slope oil if construction were to begin immediately on a Trans-Canadian Pipeline (TCP) having an accelerated throughput. Therefore, it does not appear likely that an excess supply problem in the case of TCP alternatives would occur. Indeed, the Midwest and the East Coast are likely to face severe shortages; hence an increasing pressure on price is likely. For that reason, the following discussion is restricted to the issue of supply and demand imbalances on the West Coast (District V).

District V Market Simulation Model

Excess Supply and Net Benefits

Here, using a nonstochastic computer simulation model,[2] I assess the significance of any possible oversupply in District V. I consider four cases of growth in demand: compound rates of growth of 0 percent, 2.5 percent, 3.5 percent, and 4.5 percent. The 2.5 percent case corresponds to the rate of growth in demand observed for the latter part of the 1960s in District V. The two higher rates approximate the observed recent growth in demand for Districts I to IV.

On the supply side, I have assumed that domestic supply grows at rates of 0 percent, 0.5 percent, and 1.3 percent per year. The middle estimate refers to recent trends in California production. The higher rate corresponds to the recent growth observed in District V when Cook Inlet and Southern Alaska production is included in the total.

Several assumptions are utilized for determining the quantity of foreign oil that would be permitted to enter District V. The first is that an excess supply from domestic sources would displace foreign oil. The second is that the present throughput of the Trans-Mountain Pipeline, which supplies oil from Alberta to the Puget Sound area (under an overland exemption), would remain outside the Mandatory Oil Import Quota Program and be treated as domestic oil. A third assumption is that any excess supply on the West Coast would be handled by reversing the flow of oil (backpumping) from the Puget Sound area via the Trans-Mountain Pipeline to Edmonton and then by the Interprovincial Pipeline to the Midwest. Fourth, it is assumed that all the present (1970) foreign imports would be permitted to remain at their present level. The final assumption is that the basis for determining the level of foreign imports to Districts I to IV would be utilized in District V. That is, imports amounting to 12.2 percent of domestic production in District V would be permitted to enter the district.

For the two TAP throughputs, table 29 shows the excess quantity of

[2] A copy of the computer program can be obtained from the author on request.

oil supplied at present prices that would result in District V in selected years under the institutional and demand-supply assumptions outlined above. TAP is projected to be operational for a full calendar year for the first time in 1976. Under the Alyeska production schedule the average TAP throughput in 1976 is estimated to be 900,000 barrels per day; under the accelerated schedule the estimated throughput would be 2 MMb/d. The estimated quantity of oil supplied in 1980 and 1985 for both throughput schedules would be 2 MMb/d.

Several important conclusions can be drawn from these simulation results. On the basis of recent rates of growth in demand (2.5 percent) and supply (1.3 percent) in District V, it is likely there would be an excess quantity of oil supplied into the 1990s if the entire throughput of TAP were brought to the West Coast. On the other hand, the oil-short Midwest and East Coast would be confronted with an immediate shortage of 2 MMb/d by 1975. Since this is the earliest possible time that the full TAP throughput could conceivably be made available to the markets east of the Rockies, the importance of a Trans-Canadian Pipeline is notable. If a TCP alternative has not been constructed by that time, a combination of increased domestic production,[3] increased foreign imports, or an increase in the price of crude oil would result. In chapter 3 I calculated the net benefits, tax revenues, and profits on the assumption that this would be the situation in the future. Therefore, those estimates are appropriate.

It is important to consider the effect of oversupply for the West Coast on the benefit estimates, which were made earlier using a static economic analysis. Since District V may be oversupplied with oil for perhaps fifteen years or more, the estimates of benefits, taxes, and profits for that area that were based on an assumption of no excess supply have been consistently overstated. The appropriate adjustment for these calculations depends on (1) the rates of growth in demand and supply of oil on the West Coast, and (2) the type of restrictions that might be placed on foreign oil in the future. The present value of benefit estimates, which were based on demand being sufficient to absorb the full TAP throughput, might be overestimated by as much as 75 percent.[4] Thus when dynamic conditions are taken into account

[3] Increased domestic production might come about through a reduction in state prorationing restrictions or through offshore production.

[4] This is based on the assumption that the full throughput of TAP is oversupplied for fifteen years and that the discount rate is 10 percent, thus reducing the present value of benefits to approximately 24 percent of those utilized in chap. 2. While this might be an outside estimate, many of the cases considered in table 29 imply that the benefit figures given for TAP would fall to between 24 percent and 50 percent of the values used previously for TAP.

TABLE 29. Trans-Alaska Pipeline: Excess Supply in District V at Present Prices under Various Assumptions about Demand and Supply

(thousands of barrels per day)

Assumption	Percentage growth in demand per yr.	Alyeska[a] 1976	Accelerated[b] 1976	Alyeska[b] 1980	Accelerated[b] 1985
1. Domestic supply is constant	0	443.0	1,543.0	1,543.0	1,543.0
Any excess supply backs out all foreign oil	2.5	131.0	1,231.0	995.7	667.0
	3.5	−5.0	1,095.0	740.7	223.4
	4.5	−147.6	952.4	462.5	−284.5
2. Domestic supply is constant	0	663.0	1,763.0	1,763.0	1,763.0
Canadian imports remain exempt from quota	2.5	351.0	1,451.0	1,215.7	887.0
	3.5	215.0	1,315.0	960.7	443.4
	4.5	72.4	1,172.4	682.5	−64.5
3. Domestic supply is constant	0	223.0	1,323.0	1,323.0	1,323.0
Trans-Mountain Pipeline is used to backpump any excess supply	2.5	0	1,011.0	775.7	447.0
to Edmonton, then Chicago	3.5	−5.0	875.0	520.7	3.4
	4.5	−147.6	732.4	242.5	−284.5
4. Domestic supply grows at the Calif. 1968–70 rate (0.5%/yr.)	0	482.6	1,582.6	1,609.7	1,644.3
Any excess supply backs out all foreign oil	2.5	170.6	1,270.6	1,062.4	768.3
	3.5	34.6	1,134.6	807.4	324.6
	4.5	−108.0	992.0	529.2	−183.2
5. Domestic supply grows at the Calif. 1968–70 rate (0.5%/yr.)	0	702.6	1,802.6	1,829.7	1,864.3
Canadian imports remain exempt from quota	2.5	390.6	1,490.6	1,282.4	988.3
	3.5	254.6	1,354.6	1,027.4	544.6
	4.5	112.0	1,212.0	749.2	36.8
6. Domestic supply grows at the Calif. 1968–70 rate (0.5%/yr.)	0	262.6	1,362.6	1,389.7	1,424.3
Trans-Mountain Pipeline is used to backpump any excess supply	2.5	0	1,050.6	842.4	548.3
to Edmonton, then Chicago	3.5	0	914.6	587.4	104.6
	4.5	−108.0	772.0	309.2	−183.2
7. Domestic supply grows at the 1968–70 rate of District V (1.3%/yr.)	0	548.1	1,648.1	1,722.8	1,821.8
Any excess supply backs out all foreign oil	2.5	236.0	1,336.0	1,175.5	945.8
	3.5	100.1	1,200.1	920.5	502.3
	4.5	−42.5	1,057.5	642.3	−5.7

8. Domestic supply grows at the 1968–70 rate of District V (1.3%/yr.)	0	768.1	1,868.1	1,942.8	2,041.8
Canadian imports remain exempt from quota	2.5	456.0	1,556.0	1,395.5	1,165.8
	3.5	320.1	1,420.1	1,140.5	722.1
	4.5	177.5	1,277.5	862.3	214.3
9. Domestic supply grows at the 1968–70 rate of District V (1.3%/yr.)	0	328.1	1,428.1	1,502.8	1,601.8
Trans-Mountain Pipeline is used to backpump any excess supply	2.5	16.0	1,116.0	955.5	725.8
to Edmonton, then Chicago	3.5	0	980.1	700.5	282.3
	4.5	−42.5	837.5	422.3	−5.8
10. Domestic supply is constant	0	900.0	2,000.0	2,000.0	2,000.0
Present (1970) foreign imports remain constant	2.5	588.0	1,688.0	1,452.7	1,124.0
	3.5	452.0	1,552.0	1,197.7	680.4
	4.5	309.4	1,409.4	919.5	172.5
11. Domestic supply grows at the Calif. 1968–70 rate (0.5%/yr.)	0	939.6	2,039.6	2,066.7	2,101.3
Present (1970) foreign imports remain constant	2.5	627.6	1,727.6	1,519.4	1,225.3
	3.5	491.6	1,591.6	1,264.4	781.6
	4.5	349.0	1,449.0	986.2	273.8
12. Domestic supply grows at the 1968–70 rate of District V (1.3%/yr.)	0	1,005.1	2,105.1	2,179.8	2,278.8
Present (1970) foreign imports remain constant	2.5	693.0	1,793.0	1,632.5	1,402.8
	3.5	557.1	1,657.1	1,377.5	959.1
	4.5	414.5	1,514.5	1,099.3	451.3
13. Domestic supply is constant	0	711.9	1,946.1	1,744.8	1,744.8
The east of the Rockies import quota formula is applied to District	2.5	399.9	1,634.0	1,197.5	868.8
V (12.2% of domestic production)	3.5	263.9	1,498.1	942.5	425.2
	4.5	121.3	1,355.5	664.3	−82.7
14. Domestic supply grows at the Calif. 1968–70 rate (0.5%/yr.)	0	756.3	2,003.0	1,819.6	1,858.4
The east of the Rockies import quota formula is applied to District	2.5	444.3	1,691.3	1,272.3	982.5
V (12.2% of domestic production)	3.5	308.4	1,555.3	1,017.3	538.8
	4.5	165.7	1,412.7	739.1	30.9
15. Domestic supply grows at the 1968–70 rate of District V (1.3%/yr.)	0	817.0	2,131.1	1,880.3	1,919.1
The east of the Rockies import quota formula is applied to District	2.5	505.0	1,819.0	1,333.0	1,043.1
V (12.2% of domestic production)	3.5	369.0	1,683.1	1,078.0	599.5
	4.5	226.4	1,540.5	798.8	91.6

NOTE: A minus sign indicates excess quantity of oil demanded.
[a] 900,000 barrels per day.
[b] 2 million barrels per day.

the estimated social value of a barrel of North Slope oil would increase significantly for a TCP alternative relative to the TAP alternative. It would therefore also make a TCP a more economic alternative at both higher costs and greater delays than those considered in chapter 3.

Profits and Taxes

Since decisions concerning the selection of the optimal market are typically made by the private sector of the economy, I have estimated the sensitivity of gross revenues to any supply and demand imbalances for each of the supply and demand simulations described above. These estimates are shown in table 30.

The following computations have been made for the Alyeska and accelerated throughput schedules: (1) the undiscounted gross revenues, under an assumed price elasticity of demand of −1, and (2) the present value of gross revenues, using a discount rate of 10 percent with assumed price elasticities of demand of −1 and −0.5. To make the calculations for the accelerated throughput schedule directly comparable to the Alyeska schedule, an adjustment must be made because the total throughput assumed for the Alyeska schedule is 16.74 billion barrels and for the accelerated schedule is 18.25 billion barrels.[5]

In all cases the assumption is that price changes would be restricted to the present price range of $2.25 (foreign) and $3.17 (domestic) per barrel. Therefore, it is conservatively assumed that any price change would not cause West Coast prices to go below the present world price or above the present domestic price.

Table 31 is a summary of the low and high gross revenue estimates in table 30. These estimates can be compared with gross revenue estimates presented in chapter 1 for the static analysis, which was based on the incorrect assumption of no excess supply on the West Coast. When the more realistic dynamic model is utilized, the total undiscounted revenue is approximately one-third of that used in the static model. A comparison of present values shows that the effect of excess supply is to make the expected present value of the more realistic dynamic gross revenues decline to between one-half and two-thirds of the incorrect static gross revenue estimates.

Comparisons of gross revenue estimates provide good indices of net profit and tax revenue streams under different supply and demand conditions. Accordingly, the net profit and tax revenue estimates discussed in chapter 1 are significantly overstated. Furthermore, since the esti-

[5] One adjustment factor would be to take 0.916 (= 16.74/18.25) of the gross revenue in all accelerated cases in table 22.

mates of comparable parameters for the TCP alternatives are not likely to be sensitive to dynamic supply and demand conditions (at least not in the downward direction), previous conclusions based on profits and tax revenues about the superiority of the TCP alternatives relative to TAP are greatly reinforced. The remainder of this chapter is a brief review of some of the alternative marketing practices that may be selected by the industry as substitutes for the profit deterioration expected if oil companies irrationally elect to enter into price competition.

Alternatives to Price Competition for Dealing with Excess Supply

In this section I analyze proposals to deal with excess supply in District V (if TAP begins operating in 1976) by transporting the oil to markets other than the West Coast. Such decision parameters as oil company profits, state tax receipts, and economic efficiency are examined. I emphasize, however, that there is no need to develop these secondary markets for North Slope oil if a TCP alternative is selected, since the Midwest and East Coast will need all the additional oil by 1975.

A New West-to-East Pipeline in the Lower Forty-Eight States

There are two plans for using the Puget Sound area as an import terminal to pipe any excess supply in District V eastward to the Midwest. The first is the possibility of replacing Canadian oil in the Trans-Mountain Pipeline. North Slope oil would then be pumped to Edmonton and from there on to the Midwest via the Interprovincial Pipeline. The second is the possibility of constructing a new U.S. pipeline from the Pacific Northwest via Puget Sound to the Midwest. The environmental consequences of terminal activity in Puget Sound, piping oil through earthquake-prone areas of the Pacific Northwest, and the possible necessity to cross national parks and forests would probably cause each of these routes to meet with considerable opposition from environmentalists.

Therefore, a southwestern U.S. pipeline route might also be considered. This route might follow present pipelines from Los Angeles to the Four Corners area in Arizona and Colorado and from there to the Gulf States, and it could probably be constructed for less money than a more northern route. But tanker costs to southern California would be

TABLE 30. Trans-Alaska Pipeline: District V Revenues under Various Assumptions about Demand and Supply

(billions of dollars)

Assumption	Percentage growth in demand per yr.	Cumulative gross revenue: Alyeska $E_p = -1$	Cumulative gross revenue: Accelerated $E_p = -1$	Present value of gross revenue at a 10% discount rate: Alyeska $E_p = -1$	Alyeska $E_p = -0.5$	Accelerated $E_p = -1$	Accelerated $E_p = -0.5$
1. Domestic supply is constant. Any excess supply backs out all foreign oil	0	14.9	18.1	6.1	6.0	8.0	8.0
	2.5	16.1	18.8	6.6	6.1	8.2	8.0
	3.5	17.9	20.3	7.2	6.6	8.8	8.2
	4.5	19.5	22.1	7.8	7.4	9.4	8.9
2. Domestic supply is constant. Canadian imports remain exempt from quota	0	14.8	18.1	6.0	6.0	8.0	8.0
	2.5	15.4	18.3	6.3	6.0	8.1	8.0
	3.5	16.8	19.5	6.8	6.1	8.5	8.1
	4.5	18.6	21.1	7.4	6.8	9.0	8.6
3. Domestic supply is constant. Trans-Mountain Pipeline is used to backpump any excess supply to Edmonton, then Chicago	0	15.1	18.1	6.2	6.0	8.0	8.0
	2.5	17.2	19.6	7.0	6.4	8.5	8.0
	3.5	19.0	21.5	7.7	7.1	9.3	8.6
	4.5	20.1	23.0	8.1	7.9	9.9	9.3
4. Domestic supply grows at the Calif. 1968–70 rate (0.5%/yr.) Any excess supply backs out all foreign oil	0	14.9	18.1	6.1	6.0	8.0	8.0
	2.5	15.8	18.6	6.5	6.1	8.2	8.0
	3.5	17.5	20.0	7.1	6.4	8.7	8.1
	4.5	19.2	21.8	7.7	7.2	9.3	8.8
5. Domestic supply grows at the Calif. 1968–70 rate (0.5%/yr.) Canadian imports remain exempt from quota	0	14.8	18.1	6.0	5.9	8.0	8.0
	2.5	15.2	18.2	6.2	6.0	8.0	8.0
	3.5	16.5	19.2	6.6	6.1	8.4	8.0
	4.5	18.2	20.8	7.3	6.7	8.9	8.4
6. Domestic supply grows at the Calif. 1968–70 rate (0.5%/yr.) Trans-Mountain Pipeline is used to backpump any excess supply to Edmonton, then Chicago	0	15.1	18.1	6.2	6.0	8.0	8.0
	2.5	16.8	19.3	6.9	6.3	8.4	8.0
	3.5	18.6	21.1	7.5	6.9	9.1	8.4
	4.5	20.0	22.7	8.0	7.8	9.8	9.1
7. Domestic supply grows at the 1968–70 rate of District V (1.3%/yr.) Any excess backs out all foreign oil	0	14.9	18.1	6.0	6.0	8.0	8.0
	2.5	15.4	18.2	6.3	6.1	8.1	8.0
	3.5	16.9	19.5	6.8	6.2	8.5	8.0
	4.5	18.7	21.2	7.5	6.9	9.1	8.5

8. Domestic supply grows at the 1968–70 rate of District V (1.3%/yr.) Canadian imports remain exempt from quota	0	14.7	18.1	6.0	5.9	8.0	8.0
	2.5	15.0	18.1	6.1	6.0	8.0	8.0
	3.5	16.0	18.8	6.5	6.0	8.2	8.0
	4.5	17.7	20.3	7.1	6.4	8.7	8.2
9. Domestic supply grows at the 1968–70 rate of District V (1.3%/yr.) Trans-Mountain Pipeline is used to backpump any excess supply to Edmonton, then Chicago	0	15.0	18.1	6.1	6.0	8.0	8.0
	2.5	16.2	18.8	6.6	6.2	8.2	8.0
	3.5	18.0	20.4	7.3	6.6	8.8	8.2
	4.5	19.6	22.3	7.9	7.5	9.6	8.9
10. Domestic supply is constant Present (1970) foreign imports remain constant	0	14.7	18.1	6.0	6.0	8.0	8.0
	2.5	14.9	18.1	6.1	6.0	8.0	8.0
	3.5	15.9	18.8	6.4	6.0	8.2	8.0
	4.5	17.5	20.2	7.0	6.3	8.7	8.3
11. Domestic supply grows at the Calif. 1968–70 rate (0.5%/yr.) Present (1970) foreign imports remain constant	0	14.7	18.1	5.9	5.9	8.0	8.0
	2.5	14.9	18.1	6.0	6.0	8.0	8.0
	3.5	15.6	18.6	6.3	6.0	8.2	8.0
	4.5	17.1	19.9	6.9	6.2	8.6	8.2
12. Domestic supply grows at the 1968–70 rate of District V (1.3%/yr.) Present (1970) foreign imports remain constant	0	14.7	18.1	5.9	5.9	8.0	8.0
	2.5	14.8	18.1	6.0	5.9	8.0	8.0
	3.5	15.3	18.4	6.2	6.0	8.1	8.0
	4.5	16.6	19.5	6.7	6.1	8.5	8.1
13. Domestic supply is constant The east of the Rockies import quota formula is applied to District V (12.2% of domestic production)	0	14.6	18.1	6.0	6.0	8.0	8.0
	2.5	15.4	18.1	6.3	6.0	8.0	8.0
	3.5	16.8	18.9	6.8	6.1	8.3	8.0
	4.5	18.6	20.4	7.4	6.8	8.8	8.3
14. Domestic supply grows at the Calif. 1968–70 rate (0.5%/yr.) The east of the Rockies import quota formula is applied to District V (12.2% of domestic production)	0	14.7	18.1	6.0	6.0	8.0	8.0
	2.5	15.2	18.1	6.2	6.0	8.0	8.0
	3.5	16.4	18.5	6.6	6.0	8.1	8.0
	4.5	18.2	19.7	7.3	6.6	8.5	8.1
15. Domestic supply grows at the 1968–70 rate of District V (1.3%/yr.) The east of the Rockies import quota formula is applied to District V (12.2% of domestic production)	0	14.7	18.1	6.0	6.0	8.0	8.0
	2.5	15.1	18.1	6.1	6.0	8.0	8.0
	3.5	16.2	18.4	6.5	6.0	8.1	8.0
	4.5	17.9	19.6	7.2	6.5	8.5	8.1

TABLE 31. Trans-Alaska Pipeline: Comparison of Gross Revenues in District V, Using Dynamic and Static Models

(billions of dollars)

	Gross revenue			
	Alyeska		Accelerated	
Model	Low	High	Low	High
A. Dynamic model[a]				
Undiscounted (price elasticity = −1)	14.6	20.1	18.1	23.0
Present value (10% discount rate)				
Price elasticity =				
−0.5	6.0	7.9	8.0	9.3
−1	6.0	8.1	8.0	9.9
B. Static model				
Undiscounted	53.1		57.8	
Present value (10% discount rate)	12.8		15.8	

[a] For the low case, the assumptions are a 0% growth in demand, a 1.3% growth in supply, and the application of the east of the Rockies import quota (12.2% of domestic production) in District V. For the high case, the assumptions are a 4.5% growth in demand, a 0% growth in supply, and the backpumping of any excess supply through the Trans-Mountain Pipeline.

higher than those to Puget Sound, and there would be extra transportation costs to bring North Slope oil from the Gulf States to the Midwest (at least 25¢ per barrel) or East Coast (at least 45¢ per barrel).

Two factors must be analyzed in order to determine the value of these three west-to-east North American pipelines for the interested parties. First, in order to calculate the net social benefits per barrel, the additional transportation costs to bring oil to eastern consumers via TAP must be compared with the costs of the more direct TCP system. The previous analysis concentrated on a comparison of the TAP and TCP alternatives, and it was implicitly assumed that no secondary markets would be selected for North Slope oil. Since this assumption is untenable in the TAP case, estimates are shown for the costs per barrel of the three west-to-east pipelines in table 32.

Note that for plan 2, if the existence of TAP is taken as given, the Midwest market might actually yield higher net social benefits than the West Coast market. This is because the additional costs are probably lower in this case than the additional benefits in the Midwest if it is assumed that some high-cost domestic oil would be displaced. If adverse environmental effects for both Canada and the United States are considered, this probably would not be true. Furthermore, it must be emphasized that such a roundabout route would be a far more costly method of supplying the Midwest or East Coast markets with

TABLE 32 Costs per Barrel in Excess of the Trans-Alaska Pipeline for West Coast-to-Midwest Pipeline Alternatives

(*cents per barrel*)

Plan	Cost savings relative to TAP to Los Angeles	Costs in excess of TAP to Los Angeles
1. Valdez-Puget Sound-Edmonton-Midwest		
Reduced Tanker Costs (35¢–21¢)	14	
Trans-Mountain Pipeline Costs		40
Interprovincial Pipeline Costs		
Present		51
Future		38
Net Additional Costs		
Present		77
Future		64
2. New Puget Sound to Midwest Pipeline		
Reduced Tanker Costs (35¢–21¢)	14	
Pipeline Cost Estimate		60
Net Additional Costs		46
3. New Southwest-Gulf State-Midwest Pipeline System		
Reduced Tanker Costs	0	
Pipeline to Gulf States		50
Transportation Costs to Midwest		25
Net Additional Costs		75

NOTE: These estimates are based on those found in Cabinet Task Force on Oil Import Control, *The Oil Import Question*, app. E. (February 1970).

North Slope oil than a direct overland TCP route. The indirect method would cost the nation real resources equal to about 50¢ to $1 per barrel.

Additional measures of the desirability—and therefore the likelihood—of these transportation systems from the West Coast to the Midwest are the effects of such plans on profits and taxes per barrel. The current price for the 26°–26.9° API North Slope crude oil (without a sulfur adjustment) is 64¢ per barrel greater in the Midwest than in Los Angeles ($3.81 – $3.17). Should North Slope crude prove to be lighter, this price difference would decline; newer refining methods, however, may make it increase. Only in the case of a new pipeline from Puget Sound to the Midwest (plan 2) would the additional prices result in any increase in the wellhead price used in the calculations for TAP to Los Angeles in chapter 1. Even then, if lighter crudes were used or sulfur adjustments were made, wellhead prices would decline. In all other cases, wellhead prices, and therefore taxes and profits per barrel, would decline by as much as 50¢ per barrel relative to the estimates for TAP to Los Angeles and more than $1 per barrel relative to the direct all-overland TCP to the Midwest (see chapter 3).

Larger profits and higher taxes from piping oil from District V to the Midwest are unlikely using current prices. The alternative to such higher-cost schemes might be for firms to enter into price competition on the West Coast. If the expected price decline from competitive underselling exceeded the additional transportation costs, such pipeline plans might be considered further by the oil industry. Thus a combination of price decreases on the West Coast and some high-cost movement of oil eastward might be selected to reduce the excess supply of oil expected on the West Coast if TAP were to begin operations in 1975. Profits and taxes would then be expected to decline relative to those presented in chapter 1, thus further increasing the relative superiority of a TCP alternative.

"Panama" Oil: Some Economic Comparisons

Edward L. Patton, the president of Alyeska, has estimated that if TAP begins operations in 1976 an oversupply of about 500,000 barrels per day would be likely in District V into 1980. He called this oversupply "Panama" oil and suggested that it might be sold to Japan[6] or transported from Valdez eastward via tankers to the U.S. East Coast. There are two versions of the latter proposal: (1) shipment by tanker from Valdez to Central America, a pipeline to the Caribbean Sea, and a tanker to eastern U.S. markets; or (2) transporting the oil from Valdez around Cape Horn to eastern U.S. markets.

At the present time the restrictions of the Jones Act are limiting factors on the selection of alternative marine transportation systems for dealing with the excess supply of oil on the West Coast. Thus the differences between domestic and foreign tanker rates in table 33 must be taken into account.

The Sale of North Slope Oil to Japan. This proposal has been discussed by several oil company executives. Each has suggested a different basis for such an arrangement. Rollin Eckis[7] has suggested that Japan may be willing to pay a premium for a relatively secure alterna-

[6] See Richard Corrigan, "Japan May Get Some Alaskan Oil: Foreign Flag Shipping of Exports Is Likely," *The National Journal*, July 31, 1971. In this article, Corrigan reports on interviews with U.S. and Japanese officials.

[7] An executive with Atlantic Richfield Company and a former president of Richfield Oil before its merger with Atlantic Refining Company. For a discussion of the Japanese market, see Rollin Eckis, "Alaska Oil in Domestic and World Markets," in George W. Rogers, ed., *Change in Alaska* (University of Alaska Press, 1970).

TABLE 33. Approximate Tanker Tariffs

(*dollars per barrel*)

	Foreign flag		
	Future[a]	Present	U.S. flag
Persian Gulf to Los Angeles	0.44	0.74	–
Valdez to Japan	0.20	0.34	–
Valdez to Los Angeles	0.11	0.19	0.35
Valdez to Panama to New York[b] (pipeline across Panama)	0.32[c]	0.53[c]	0.65 to 1.06[d]
Valdez to New York around Cape Horn[b]	0.66	1.11	1.32 to 2.22[d]
Seattle to Los Angeles	–	–	0.15

SOURCE: Based on rates in Cabinet Task Force on Oil Import Control, *The Oil Import Question* (February 1970), and a simplified linear approximation.

[a] Estimates based on economies due to supertankers.

[b] Perhaps 10¢ per barrel should be added to these estimates if the oil is first brought to the Virgin Islands.

[c] Terminal and pipeline costs must be added. Assuming 25¢ for construction and operating costs, I use estimates of 57¢ and 78¢.

[d] Assuming U.S. flagships would cost twice as much per barrel for operating and construction costs.

tive to buying oil in the Middle East, which supplied more than 80 percent of the rapidly growing Japanese demand for oil.

John M. Houchin, deputy chairman of Phillips Petroleum, proposed a two-tier pricing system for marketing North Slope oil in Japan.[8] His proposal, known as the import-for-export program, included a recommendation to urge the President of the United States to approve the right of the exporting company to import a barrel of oil on the East Coast for every barrel exported from the North Slope. Because the United States would be producing an amount of oil equivalent to the increase in imports, proponents of the program averred that national security and balance of payment effects would be mutually offsetting and therefore negligible.

Edward Patton indicated that as much as 100,000 barrels per day may be sold to Japan in 1980. If Cook Inlet production, which is being used to supply natural gas to Tokyo, is displaced in District V markets by North Slope oil, the total Alaskan oil going to Japan in 1980 may well exceed 300,000 barrels per day.

[8] *Oil Import Controls*, Hearings before the Subcommittee on Mines and Minerals, House Interior and Insular Affairs Committee, 91 Cong. 2 sess. (1970). At the present time, Phillips is not a holder of major proved-up reserves on the North Slope. It does have reserves in Cook Inlet, however, that are both closer to Japan and more likely to be backed out of District V by North Slope oil.

Table 33 gives estimated tanker costs of approximately 20¢ to 34¢ per barrel for marine transportation from Valdez to Japan in foreign flagships. In 1970 the average price of crude oil in Japan was about $1.83 per barrel. This price might be expected to increase to slightly more than $2 by 1975.[9] Therefore, North Slope oil transported via a TAP-tanker system to Japan would be worth approximately $1.15 per barrel less at the wellhead than if it were sold in Los Angeles in 1975 at the current domestic price of $3.17 per barrel.

Marketing oil in Japan would yield a before-tax profit of approximately 95¢ per barrel.[10] If the same barrel of oil were sold in Los Angeles, the profit before taxes would be approximately $2 per barrel—if oversupply does not cause prices to fall. A combination of falling West Coast prices, appreciating world prices, and possible Japanese willingness to pay a national security premium might make up these differences.

From the standpoint of social benefits to the United States the Japanese alternative would be inferior to both TAP and TCP. Furthermore, if the environmental damages in the Arctic and the real costs in the oil-short Midwest and West Coast due to increased dependence on foreign oil are taken into account, the size of the net social costs of this Japanese alternative increase significantly. These conclusions apply more or less equally to society, the treasury of Alaska, and the oil companies when oil is sold to Japan without any offsetting compensation. They would not hold if an import-for-export plan were approved. Under such a plan foreign oil would be imported to the East Coast outside the Mandatory Oil Import Quota Program to compensate the company that exported the Alaskan oil. This would mean a windfall profit of approximately $1.75 per barrel, based on a price of approximately $4.06 per barrel (26°–26.9° API) and a cost of about $2.30 for an imported barrel.

In table 34 these results are used to estimate the effect on profitability. If the net profits before taxes from selling North Slope oil in Japan (about 95¢ per barrel), the savings in taxes paid to the state of Alaska (about 23¢ per barrel, based on 20 percent of the wellhead price difference between selling oil in Japan at $2 or in Los Angeles at $3.17 per barrel), and the economic profit ($1.75) from using foreign oil on the East Coast are summed, the net revenue or value to the oil companies amounts to nearly $3 per barrel ($2.93 = $0.95 + $0.23 + $1.75)

[9] The $1.83 per barrel is based on averages for the first quarter of 1970 found in *Platt's Oilgram Price Service*, May 10, 1971.

[10] Calculated by subtracting production costs (25¢), transportation costs for a pipeline (60¢), and tanker costs (20¢) from a price of $2 per barrel.

TABLE 34. Net Revenue Effects of an Import-for-Export Program under Present Prices

(*dollars per barrel*)

	Los Angeles	Japan
Posted prices	3.17	2.00
Approximate production costs for North Slope oil	0.25	0.25
Approximate pipeline costs	0.60	0.60
Approximate tanker costs	0.35	0.20
Before-tax new revenue	1.97	0.95
Bonus from importing foreign crude oil on the East Coast outside the Mandatory Oil Import Quota Program		1.75
Savings in taxes to the state of Alaska based on a wellhead price difference of $1.17 per barrel		0.23
Per barrel equivalent net revenue	1.97	2.93
Difference	0.96	

for each barrel of North Slope oil that is marketed in this manner. While the oil companies would find these profits approximately $1 per barrel higher under this arrangement, the state of Alaska would receive less than half as much per barrel in royalty and severance taxes. Furthermore, the social costs of producing a single barrel of oil for U.S. consumers would equal the sum of the cost of imported foreign oil (approximately $2.30) and the cost of supplying Japan (approximately $1.05). While national security and benefits from trade may also be present, this estimate of the costs of a barrel of oil exceeds all previous North Slope estimates by about $2 per barrel.[11]

The Sale of North Slope Oil on the U.S. East Coast. While the movement of Alaskan oil to Japan in foreign flagships is permitted, the transportation of oil to the U.S. East Coast from Alaska in foreign tankers is not. There are two possible methods of circumventing this restriction. First, oil shipped to the eastern part of the United States via a new Central American pipeline might be considered outside the restrictions of the Jones Act, since two foreign ports and a foreign pipeline would be involved.

A second exception to the Jones Act is more likely. It would occur if North Slope oil were transported via a Central American pipeline or around Cape Horn to the Virgin Islands, where it could be refined and

[11] If the imported oil is produced by the same company that exports Alaskan oil, the royalties paid to the producing country may be credited against U.S. corporate income taxes. Such a tax advantage has not been included in the analysis of benefits and costs, since it is a transfer payment. It would affect profitability, however, and must therefore be included when private decisions are analyzed. This is especially true when the company producing North Slope oil would also be the producer of foreign oil that would enter under an import-for-export program.

sent to markets in the eastern part of the United States. Amerada Hess Corporation is reported to have expanded its refinery complex in the Virgin Islands to 450,000 barrels per day in order to process an expected large quantity of North Slope oil for marketing in the United States.[12]

The Virgin Islands represent something of an anomaly in the manner in which oil restrictions are applied. First, the islands are exempt from the restrictions of the Jones Act. Second, foreign oil may be imported into the islands for refining. But the amount of oil that is permitted into the United States is restricted under the present Mandatory Oil Import Quota Program.

No appreciable amount of domestic oil has been sent to the Virgin Islands to be refined and then sent back to the U.S. mainland for consumption. Therefore there are no precedents for shipping Alaskan oil to the Virgin Islands and simultaneously circumventing the Jones Act and the Mandatory Oil Import Quota Program. But the proponents of this alternative, unlike those of the import-for-export program, would probably not need to seek executive approval.

From both an economic and a financial standpoint, several calculations can be made to analyze these Panama alternatives. Table 35 shows that, first, a Central American pipeline would probably be a more efficient expenditure of resources, both social and private, than a similar Cape Horn alternative that would cost about 44¢ (foreign flag) to 92¢ (U.S. flag) more per barrel. Second, excluding any benefits to the nation gained from shipping goods in domestic bottoms, the cost differentials between foreign and domestic tankers indicate that foreign flagships are more efficient. A conclusion based solely on cost differences, however, must be carefully interpreted because of the intentions of the Jones Act.

Such questions aside, the estimated average costs for North Slope oil shipped directly to New York in U.S. flagships would be approximately $1.95 per barrel and $2.62 per barrel, respectively, for the Central American pipeline and Cape Horn alternatives. These cost estimates are well in excess of the total cost estimates in chapter 3 of $1.32 to $1.71 per barrel at a 10 percent discount rate for a direct TCP extended to New York.[13] Therefore, these extra costs in transportation would represent real losses of resources for the nation if this roundabout

[12] See Corrigan, "Japan May Get Some Alaskan Oil."

[13] Note that the values shown in chap. 3 for transporting oil to Chicago must be increased by two factors to determine these costs: production costs of 25¢ per barrel and 25¢ per barrel for transportation between Chicago and New York.

TABLE 35. Costs and Revenues under Different Plans to Transport North Slope Oil to the East Coast via the Trans-Alaska Pipeline and Tankers

(dollars per barrel)

	Valdez-Central America-Virgin Islands	Valdez-Cape Horn-Virgin Islands	Valdez-Central America-New York	Valdez-Cape Horn-New York
Approximate costs				
Production	0.25	0.25	0.25	0.25
Pipeline (TAP)	0.60	0.60	0.60	0.60
Tanker costs[a]				
Foreign flag	0.54	0.98	0.44[b]	0.88[b]
U.S. flag	0.95[b]	1.87[b]	0.85	1.77
Central American pipeline	0.25	—	0.25	—
Total cost				
Foreign flag	1.64	1.83	1.54[b]	1.73[b]
U.S. flag	2.05[b]	2.72[b]	1.95	2.62
Approximate present prices (26°–26.9° API)				
New York price	4.06	4.06	4.06	4.06
World price	2.01	2.01	2.01	2.01
Net revenue before taxes				
New York price				
Foreign flag	2.42	2.23	2.52[b]	2.33[b]
U.S. flag	2.01[b]	1.34[b]	2.11	1.44
World price				
Foreign flag	0.37	0.18	0.47[b]	0.28[b]
U.S. flag	0.04[b]	0.71[b]	0.06	0.61

[a] Tanker costs are based on a straight averaging of the range shown in table 33 for each case. An educated guess is that there would be an additional 10¢ for handling in the Virgin Island cases.

[b] This estimate is the least likely of the two shown, based on the most probable interpretation of the Jones Act.

method of supplying the East Coast were selected to deal with the excess supply of oil on the West Coast. Furthermore, the difference between the costs of supplying the Midwest through New York (an even more roundabout plan) and a direct TCP, are approximately 94¢ to $1.61 per barrel when the average cost of $1.26 per barrel at a 10 percent discount of a direct pipeline to the Midwest is used as a basis for comparison.

If the crude oil were shipped to the Virgin Islands for refining, additional transportation costs would be expected, but the prospect of using lower-cost foreign tankers would be greater. The expected savings in resources to be gained from using foreign tankers would more than compensate for the additional distance and handling costs required.

Table 35 shows that such savings would range from 41¢ to 89¢ per barrel for the Central American and Cape Horn routes respectively. Therefore, the socially uneconomic decision to divert oil to the Virgin Islands would probably be selected if the restrictions of the Jones Act could be circumvented only by this action.

All these plans would be acceptable to the oil companies only if the domestic price rather than the world price were expected on the East Coast. At the same time, in the Virgin Islands examples, if the world price were used to determine wellhead prices in Alaska, the higher costs (even in the foreign flagship case) would result in royalty and severance taxes to the state of Alaska of only about 10¢ per barrel. This is about 35¢ per barrel less than the expected royalty and severance tax if that same oil were sold in California at the current price. If, after refining, the oil were then permitted to enter the East Coast market outside the import quota system, this saving in taxes would represent additional profits to the oil companies with no accompanying loss in revenue. The state of Alaska could protect itself from both the Japanese and Virgin Islands marketing plans by passing a law similar to those in Middle East countries, which use a posted price for calculating taxes and netback, regardless of where the oil is marketed. Such decisions are transfers between oil companies and the state and indicate an equity issue but will not affect economic efficiency.

One case is particularly interesting to consider: oil would be transported in foreign tankers (outside the Jones Act) from Valdez to Central America, piped across Central America and carried in foreign tankers to the Virgin Islands, sold at the world price, and refined and sent to the East Coast in addition to the import quota levels. This plan would yield profits before taxes of $2.42 per barrel, with an additional tax savings of about 35¢ per barrel. Equivalent oil company profits would be about $2.77 per barrel. At the same time, state taxes would be their lowest per barrel, and the additional social costs of this plan would be about 13¢ to 32¢ per barrel greater than the average of those of TCP direct pipeline transportation systems.[14]

Undoubtedly marketing North Slope oil in either Japan or on the U.S. East Coast via the Virgin Islands would be profitable. Generally, however, profits, taxes, and net benefits would be lower than the esti-

[14] This is based on an average total cost on the East Coast of a TCP alternative of $1.51 per barrel compared with $1.64 per barrel for a Central American system and $1.83 per barrel for a Cape Horn system. If the Midwest were the principal North Slope market, these extra costs would be approximately 63¢ and 82¢ respectively.

mates above if it were assumed that North Slope oil would be sold on the U.S. West Coast at the current price. Thus, the values used for these parameters, when compared with the TAP and TCP alternatives, should be adjusted downward in the case of TAP. This adjustment increases the relative advantages of a direct North Slope-to-Midwest transportation system, even if this system were to be delayed further. An import-for-export plan or a Virgin Islands plan would result in lower taxes and social benefits, but profits would increase to their greatest level—nearly $3 per barrel before taxes.

The realistic issues considered in this chapter greatly strengthen the conclusion that there are strong economic, financial, and environmental reasons for selecting a TCP route to the U.S. Midwest and East Coast rather than the combined TAP-tanker system. When supply and demand imbalances on the West Coast are taken into account, the net benefits of a TCP alternative increase relative to TAP.

The possibility of using either the Japanese or Virgin Islands plan for handling excess supply on the West Coast accounts in large measure for the apparent irrationality of the oil companies' selection of the TAP route rather than a more profitable TCP alternative. Furthermore, while the cost of potential delays in constructing the TCP alternative is relatively unimportant at discount rates of 10 percent or 12 percent, these delays may be considered quite costly by oil companies that may use a much higher discount when making decisions about the internal rate of return. Rates of return in the 30 percent range[15] coupled with such possible profitable alternatives as the import-for-export and Virgin Islands plans—as well as the corporate merger agreements involving British Petroleum and SOHIO—may totally account for the private decision to select the socially inferior TAP alternative.

Private profit-maximizing decisions of this kind are the essence of a freely functioning, perfectly competitive economy. But when market restrictions like those described above are present, concentrating on profitability may not produce economically efficient results. Similarly, state tax revenues may not always be positively associated with the profitability of an alternative. If adverse environmental effects are also present but not included in profit and loss statements, it is inappropriate for society (or its elected representatives) to abrogate responsibility for examining the various route and marketing alternatives.

[15] The average rate of return on actual income received by domestic oil companies in 1970 was about 40 percent as calculated from data presented in American Petroleum Institute, *Joint Association Survey of the U.S. Oil and Gas Producing Industry, 1970* (Washington, D.C., November 1971).

6

The Case Against TAP: Conclusions of the Analysis Summarized

As the environmentalists' case against the Trans-Alaska Pipeline (TAP) slowly proceeds through the courts to ultimate settlement in the Supreme Court, Alaskan officials and oil men on the North Slope look to their ledgers in dismay. With construction of the controversial project halted, oil companies see an investment of more than $1 billion "standing still," and Alaskans must do without an expected $300 million a year in royalties and taxes. Their losses undoubtedly are temporary.

One of the firmest conclusions of my study is that the development of North Slope oil will bring substantial economic benefits to the nation as a whole. Assuming—as the U.S. Department of the Interior (USDI) has implicitly assumed—(1) that foreign crude oil would be used on the U.S. West Coast in lieu of North Slope oil, (2) that all North Slope oil is needed at present on the West Coast, and (3) that no superior transportation and marketing alternatives exist, the present economic value of TAP to the nation, at a 10 percent discount rate, is about $3 billion to $6 billion, or almost $15 billion undiscounted. Under such circumstances, a decision to totally forgo development of North Slope oil would mean that the nation judged the environmental damage to Alaska and to West Coast waterways and ports to be greater than or equal to this amount.

The crux of the legal case, however, is not *whether* North Slope oil should be allowed to flow into markets, but *how* and *when*. TAP would slice though a fragile Arctic ecosystem to the port of Valdez, where tankers would pick up the oil for delivery to the West Coast. Aroused by this threat to marine life and a magnificent wilderness area, environmentalists based their suit against USDI on failure to comply with the legal requirement to produce a detailed statement of how TAP could affect the quality of the environment and of alternatives to the route. Many of their fears concerning ecological damage were confirmed by the environmental impact statement later submitted by USDI in the spring of 1972. On the basis of this admittedly frank report, the TAP-tanker system can hardly be regarded as the most benign means of transporting Alaskan oil. Far less destructive routes—if longer—turned up in the alternatives presented by USDI.

My own analysis of these and other alternatives shows the TAP-tanker system to be inferior environmentally to a completely overland pipeline from the North Slope to the U.S. Midwest and East Coast. This conclusion is shared by a number of experts who have asserted the superiority of a Trans-Canada route, principally because it would avoid the threat of earthquakes and avalanches in southern Alaska and, lacking tanker operations, would make no contribution to marine pollution. Indeed, USDI conceded that, in view of the almost certain prospect of a gas pipeline through Canada to deliver North Slope natural gas, a joint oil and gas transportation system through Canada would be environmentally superior in all respects to a separate line through Alaska and a gas line through Canada.

Specifically, the route that I labeled TCP-I could utilize a natural corridor and would be hampered far less by formidable physiography than TAP. TCP-IV, identical to the TAP route for a considerable distance, would have the advantage of a man-made corridor in Canada, and an existing road would simplify connecting the Canadian oil provinces to this route.

Then, why TAP? Because it appeared to be the quickest and most profitable way to get North Slope oil on the market?

The legal injunction against TAP not only forced a careful consideration of the environmental effects of the pipeline, but also directed attention to the need for a more searching analysis of the economics of North Slope oil development. My study is an attempt to help meet that need, and my somewhat unexpected finding is that TAP would *not* be the most profitable transportation system for Alaskan oil. Economically, a TCP supplying the higher-valued petroleum markets of the Midwest and East Coast would be superior to TAP for the nation, for Alaska, and—except under certain interesting schemes for the ultimate delivery of oil—for the oil companies themselves.

Higher prices without a similar increase in costs usually mean greater profits. Therefore, I find that a decision to permit the oil companies to carry out their stated intent to produce North Slope oil for sale exclusively on the West Coast—using TAP and forgoing a TCP—would cost them about $3 billion in present value of profits (at a 10 percent discount rate) or about $16 billion undiscounted. Oil companies, however, do not make a practice of passing up such attractive profits. Their apparently irrational choice has been made on the basis of their own view of how to realize the most from North Slope oil.

The Profit Motives behind the Choice of TAP

Although the price of oil is higher east of the Rockies—the area that would be serviced by a TCP—than on the West Coast, the oil companies recognized profit potentials in the TAP-tanker system that made it more desirable to them than a TCP. One of these potentials lies in Japan. The other could be realized through a roundabout marine extension of TAP to the higher-priced East Coast market and would constitute a means of skirting institutional restrictions on transporting and marketing oil.

If TAP were built, if the plan to ship Alaskan oil directly to the West Coast were carried out, and if import restrictions similar to those in the rest of the nation were imposed on the West Coast, the probable result would be an excess supply of oil on the coast that could average as much as 1 million barrels a day as late as 1985 or 1990. The oil interests have proposed an import-for-export plan that would make marketing this excess in Japan a very profitable enterprise.

Under this plan, excess oil would be shipped in foreign tankers from Valdez to Japan. The oil companies would pay lower royalties to Alaska and save on transportation costs. Moreover, they would avoid the restrictions of the Jones Act, which requires that U.S. ships must be used for cargo transported between domestic ports. The exporting company would be allowed to increase its imports of foreign crude oil on the East Coast—again, in less expensive foreign tankers—for sale at the highest possible prices in the world.

Proponents of this plan argue that it would satisfy the national security and the balance of payments requirements of the Mandatory Oil Import Quota Program. But, if approved, these East Coast imports would be in addition to the present permissible level of foreign imports and would reap extraordinary profits because of the two-part pricing scheme involved in the import-for-export plan.

A second plan calls for the use of foreign tankers to transport North Slope crude oil to Central America, where a new pipeline would be built to pump the oil to another fleet of foreign tankers that would carry oil to the Virgin Islands for refining. By selling oil to themselves or engaging in "swaps" at the world price, the companies would reduce the wellhead price and therefore the taxes to the state of Alaska. After refining, this domestic crude might then be shipped to the East Coast outside the Mandatory Oil Import Quota Program, reaping larger profits than a direct overland pipeline through Canada.

In addition to the concern over profits that prompted these ingenious and complex alternatives, the oil companies undoubtedly are apprehensive over certain other aspects of an overland route across Canada. One fear is that the construction of an international pipeline would be accompanied by political jockeying that would cause endless delays. By selecting an Alaskan route the oil companies minimize the number of political entities with which they must contend. The U.S. federal government and the state of Alaska naturally would be involved in any development project. But the completely overland routes would bring in the Canadian government and that of the various provinces as well. Fear that the Canadian government will insist on a guaranteed share in the throughput capacity of a TCP is also present. Either of these occurrences could significantly alter the financial returns to the companies.

For example, the average rate of return for U.S. oil and gas production before corporate income taxes was about 40 percent in 1970. Furthermore, special tax treatment reduces taxes paid on oil production, which meant that in 1968 the ten largest domestic oil companies paid average corporate income taxes of less than 10 percent. Therefore, while a delay of a year in producing North Slope oil might cost the nation or the state of Alaska about 10¢ per dollar forgone per year, the oil companies on the North Slope find their opportunity costs to be several times greater than this amount. In addition, British Petroleum (BP) has merged with the Standard Oil Company of Ohio on the basis of an agreement that BP must produce 600,000 barrels of oil per day by the end of 1977. The possibility of delays and of the loss of pipeline throughput control plus the possible displacement of some Alaskan oil with Canadian oil if a TCP is built helps to explain to a large extent why BP has selected and pursued the TAP alternative.

Factors in Favor of a TCP Alternative

The chief characteristic of the industry point of view is self interest. Undeniably, implementation of the oil companies' plans to bypass industry regulations—and I grant these are often arbitrary—would enhance the profitability of TAP over a TCP alternative. But my analysis clearly shows that profit gained for the oil companies in this manner would be at the expense of oil consumers, taxpayers, the U.S. maritime industry, and the state of Alaska.[1]

Likewise, the companies' decision in favor of the route they judge to be the fastest is understandable, but it is not in the larger community interest. Possible delays in order to build a Canadian pipeline may not be economical for oil companies, which would have to forgo their accustomed rates of return, but a TCP would result in greater benefits for the nation than would TAP.

Furthermore, sharing a TCP throughput capacity with Canada has points in its favor. According to my analysis, sharing may be necessary in order to meet production schedules to supply domestic needs. In addition, TCP-I would pass through likely areas of Canadian oil discoveries. Since the development of Canadian oil and gas depends on an open market in states east of the Rockies, the pipeline route that is chosen should be the one that is most beneficial to the environment and economies of both the United States and Canada. Although corporate merger agreements might be detrimentally affected if Canada shared in the throughput of a TCP, the net social benefits and the present value of Alaskan tax revenues would still favor a TCP route over TAP.

Although the greater length of a TCP would probably result in higher construction costs than those of TAP, a TCP would not require the port and terminal facilities and tankers needed for the TAP system. Thus, it is not clear that the overall costs of a Canadian oil line would be more. Actually, some early estimates made by oil companies indicated that the total costs of the two routes would be nearly equal. Most current estimates show the total difference to be less than $1 billion if either route were to be constructed at the same point in time.

But there are broader economic considerations than construction costs. How, for example, would the oversupply of oil expected on the West Coast if TAP were built affect national net economic benefits?

First, it is important to remember that, unlike the East Coast, the

[1] By passing legislation setting a posted wellhead price for tax purposes, the state of Alaska can protect itself to some extent from the ingenious marketing schemes described above.

states east of the Rockies face probable oil and natural gas shortages. Handling the excess supply by building new pipelines to transport oil eastward into other states would increase both environmental and economic costs. Since taxes levied by the Oil Petroleum Exporting Countries (OPEC) are expected to increase, the value of domestic oil to the nation will be greater in the future; therefore it would be inefficient for the nation to produce U.S. oil for export before it is actually needed domestically. Because domestic markets in the Midwest would need the entire expected Alaskan production, a TCP would result in higher taxes paid to Alaska and greater economic benefits to the nation.

The price of oil in Chicago ($3.65–$3.81 per barrel) is greater than the price of oil similar in quality to North Slope crude in Los Angeles ($3.17 per barrel). These price differences far exceed any additional cost estimates of a TCP route. Therefore, Alaska—under present royalty and severance tax provisions—would find that if either route could be built without delay the present value of Alaskan tax revenues would be about $1 billion greater with a Canadian pipeline. Undiscounted this would equal a loss of about $5 billion.

Second, because of institutional restrictions and price differences, given the absence of Alaskan oil it would be more unlikely to expect the eastern states to be supplied with foreign oil than the West Coast. If foreign crude were restricted as an alternative source of supply east of the Rockies and higher-cost domestic crudes were used to supply consumers in lieu of North Slope oil, the economic efficiency benefits to the nation would increase significantly for a TCP relative to TAP. These extra benefits would amount to a present value of about $5 billion, discounted at 10 percent, and a cumulative total over the undiscounted life of the project of $15 billion. They are so great that a Canadian line would be economically superior, even if it were delayed two or three years and circumstances caused the costs of a Canadian alternative to increase by 20 percent.

In short, TAP not only would direct oil to what analysis reveals to be a poor choice of market, but from beginning to end would require new construction. A TCP, on the other hand, would supply oil to U.S. markets most in need of it—and perhaps at much lower capital costs because part of the system or rights-of-way for it already exist. Existing pipelines lying along either the TCP-I or TCP-IV routes mean that the new lines could utilize the corridors of the old ones. The Interprovincial Pipeline, which carries Canadian oil from Edmonton to eastern Canada and the U.S. Midwest, could be used as part of a TCP to carry North Slope oil to the Midwest. Extending the route beyond the

Chicago market would not affect profits and wellhead prices because price differences between the Midwest and East Coast markets equal transportation cost differences. Furthermore, the cost savings and benefits associated with a multipurpose (oil-natural gas) corridor would greatly increase the social benefits of a TCP route.

It is not surprising that the possibility of a Trans-Canadian pipeline for Alaskan oil has never been thoroughly explored with the Canadian government—although there is every sign that Canadian officials are receptive to such a proposal. After all, this alternative has been resisted by both the oil men and USDI and has suffered in comparisons with TAP because too little has been known about its economic and environmental merits.

No one analyst can produce all the answers needed to resolve the controversy over Alaskan oil, particularly when so many variables and unquantifiable factors are involved. The limitations of my own study have been pointed out in the preceding chapters. But it is my hope that the merits of TCP alternatives will receive further consideration before a final—perhaps irreversible—decision ends the controversy over TAP. The consequences of a TAP-tanker system for transporting Alaskan oil could haunt an entire continent for many years to come. Even USDI acknowledged this in its environmental impact statement: "Because of the scale and nature of the project, the impact would occur on abiotic, biotic, and socioeconomic components of the human environment far beyond the relatively small part . . . of Alaska that would be occupied by the pipeline and oilfield."[2]

[2] U.S. Department of the Interior, *Final Environmental Impact Statement Proposed Trans-Alaska Pipeline*, vol. 1, *Introduction and Summary*, "Summary Sheet" (1972), p. 1.

APPENDIX A

Recent Trends in the U.S. Crude Oil Industry

At the present time (1972) the United States operates under a Mandatory Oil Import Quota Program that restricts the amount of foreign oil that can be used in the United States. The program operates differently in the states west of the Rocky Mountains—Petroleum Administration for Defense (PAD) District V[1]—than it does in the states east of the Rocky Mountains (PAD Districts I–IV). These districts are shown on map A-1.

Market Structures

District V

In PAD District V domestic production is fully protected from foreign competition at an agreed upon domestic price in these states. If the quantity of oil demanded exceeds the quantity of oil supplied by domestic and Canadian producers, then foreign oil has been permitted to enter District V to meet the excess quantity demanded at a historical price. It should also be pointed out that Canadian producers are somewhat restricted by the existing throughput capacity of the Trans-Mountain Pipeline that transports oil from Edmonton to Puget Sound. In addition, interdistrict flows of oil have been negligible.

[1] PAD District V includes Arizona, Nevada, California, Oregon, Washington, Alaska, and Hawaii.

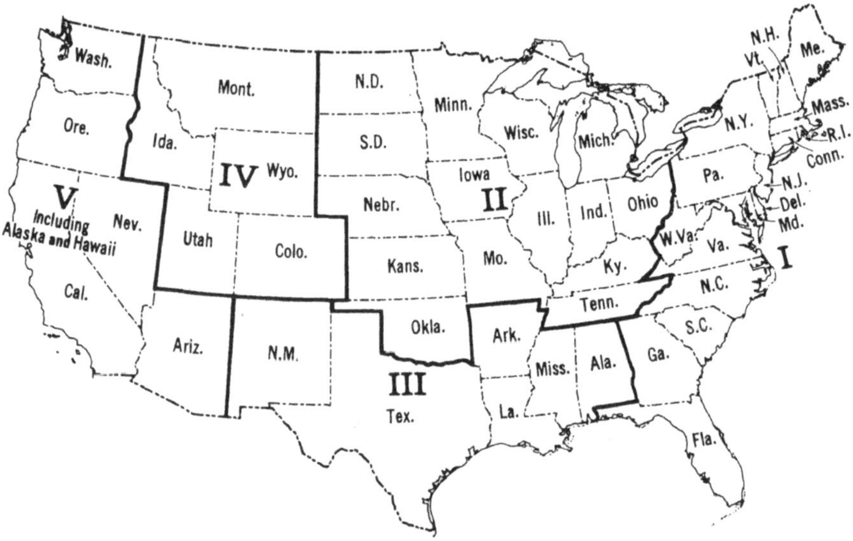

MAP A-1. Petroleum Administration for Defense (PAD) districts. *Source*: Bureau of Mines, "Crude Petroleum, Petroleum Products, and Natural-Gas-Liquids: 1969 (Final Summary)," *Mineral Industry Surveys*, December 15, 1970.

Figure A-1 depicts diagrammatically the effect of these market restrictions on the supply function and the equilibrium price in District V. Let S_d be the marginal cost curve of domestic producers in District V and assume for simplicity that Canadian producers are treated as domestic producers. At a price of \$3.17 per barrel (using 26°–26.9° API for reference purposes) the quantity of oil demanded exceeds the quantity of oil supplied by an amount equal to Q_F. This is the amount of foreign oil that is permitted to be imported into District V. Even if the marginal cost (for simplicity, I assume that the marginal and average costs of foreign oil are equal)[2] of foreign oil in amounts that exceed Q_F are lower than domestic costs, the level of foreign oil is restricted to Q_F. Therefore, the supply curve for foreign oil (S_F^v) is perfectly inelastic for amounts in excess of Q_F.

If domestic sources of supply could produce more than the amount shown in figure A-1 and all foreign oil were prohibited, the higher price and lower quantity associated with point A would be the new equilibrium position. Alternatively, if all foreign oil had costs repre-

[2] A large proportion of the cost of foreign oil is tanker costs and royalty taxes. Since these costs are generally constant, the assumption made in the text is probably true.

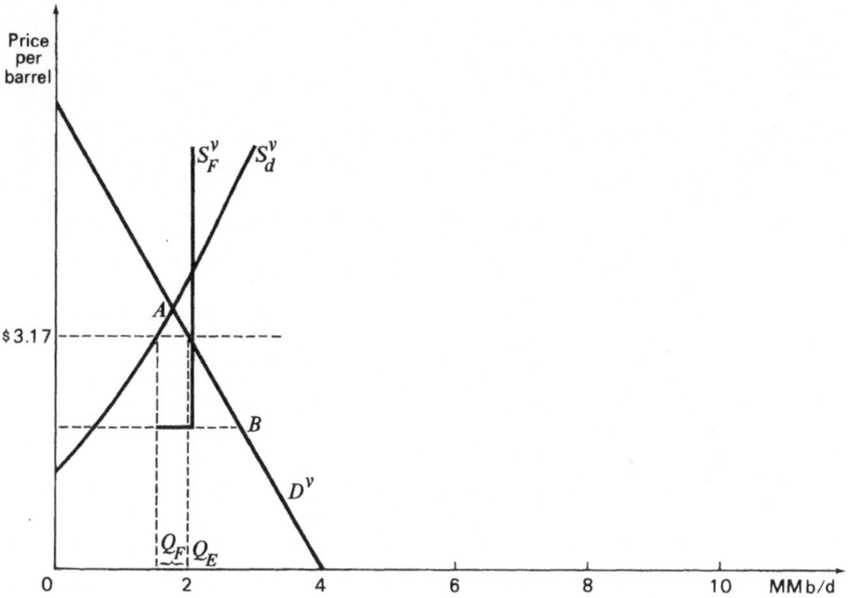

FIGURE A-1. Present market conditions in District V.

sented by the completely elastic portion of the S_F^v curve, the new equilibrium point would be at *B*, and quantity would increase and price would decline. This free market situation would result in a decline in domestic production, even if the total consumption increased, since much of it would not be competitive at the world price (which for the purposes of this analysis is equal to production and transportation costs plus taxes paid to the producing country).

Figure A-2 shows the resultant aggregate supply curve when the present import restrictions are retained. This curve is derived by summing S_F^v and S_d^v horizontally and is designated S_{d+F}^v. Assume that, at a price of $3.17 per barrel, demand grows from the equilibrium level of Q_E^{71} to Q_E^t. For simplicity also assume that $(Q_E^t - Q_E^{71})$ equals Q_{TAP}, the expected TAP throughput. This is shown by the shift in the demand curve from D_o^v to D_t^v. If foreign oil were not expanded to meet this additional demand, a price increase would be expected along with a lower equilibrium quantity than Q_E^t. This is shown by point *C*.

If the Trans-Alaska Pipeline (TAP) were built with an assumed constant marginal cost (shown in figure A-2 by segment *AG*) and its throughput rate were equal to Q_{TAP}, several increased benefits would result. First, there would be an increase in the quantity of oil con-

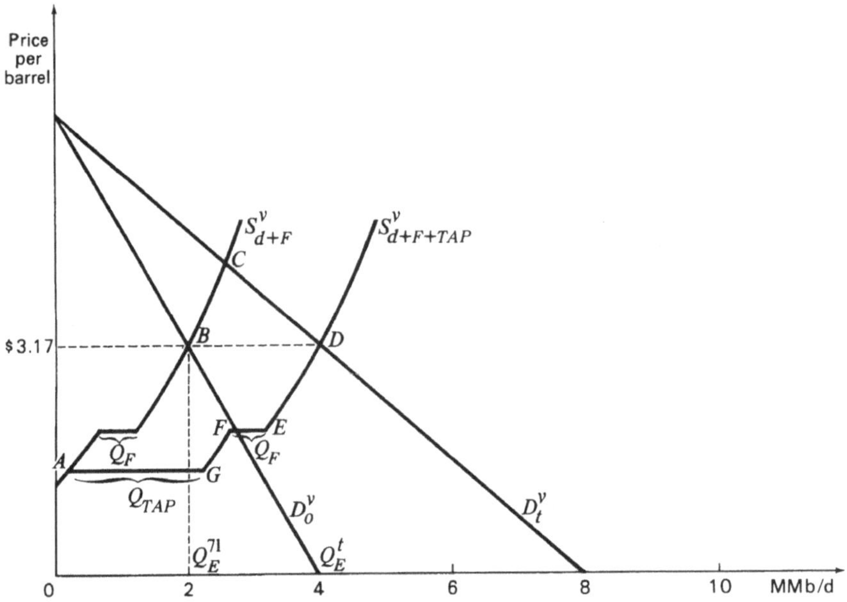

FIGURE A-2. Market conditions in District V with and without the Trans-Alaska Pipeline.

sumed and a decline in price as represented by the movement from point C to D. Second, an additional measure of the benefits would be present if the oil delivered by TAP were lower in cost than alternative sources of supply, whether foreign, domestic, or some combination of foreign and domestic.

The total economic benefits of TAP can be shown for the two assumptions used in figure A-2. These assumptions are that (1) demand grows from Q_E^{71} to Q_E^t at a price of $3.17 per barrel, an amount that equals the throughput of TAP; and (2) foreign oil is restricted to its present level, Q_F. The irregular area *ABCDEFG* would represent the total benefits that could be attributed to TAP and North Slope oil production before external effects, such as environmental damages or forgone opportunity costs, are considered.

East of the Rockies

The supply curve of crude oil east of the Rockies, that is, PAD Districts I to IV, is derived somewhat differently. First, the amount of crude oil and product imports (except residual fuel oil) is restricted to 12.2 percent of domestic production. Second, Canadian crude is treated as an import and must be subtracted from the total oil import allocations. Third, domestic oil is restricted by a system of state demand prorationing restrictions.

The effect of the domestic restrictions is that the individual marginal cost curves of each producing field, well, or firm cannot simply be added horizontally. Instead, each producing unit is assigned a certain allowable level of production per month. The result is that some low-cost wells are sometimes idle, while higher-cost wells are producing. Firms will produce from each well the allowable quantity of oil as long as the marginal costs of a barrel of oil are less than the market price. Summing such marginal cost curves horizontally yields an aggregate domestic supply curve that will be on or above the market supply curve that would occur if prorationing restrictions were removed.

The resulting market equilibrium means a loss in economic efficiency, a higher price paid by consumers, and a lower quantity of oil consumed in each time period. It is averred that these deleterious economic effects are offset because more domestic oil is ultimately recovered than would otherwise be the case.

In figure A-3 I show the supply curve for Districts I and II that

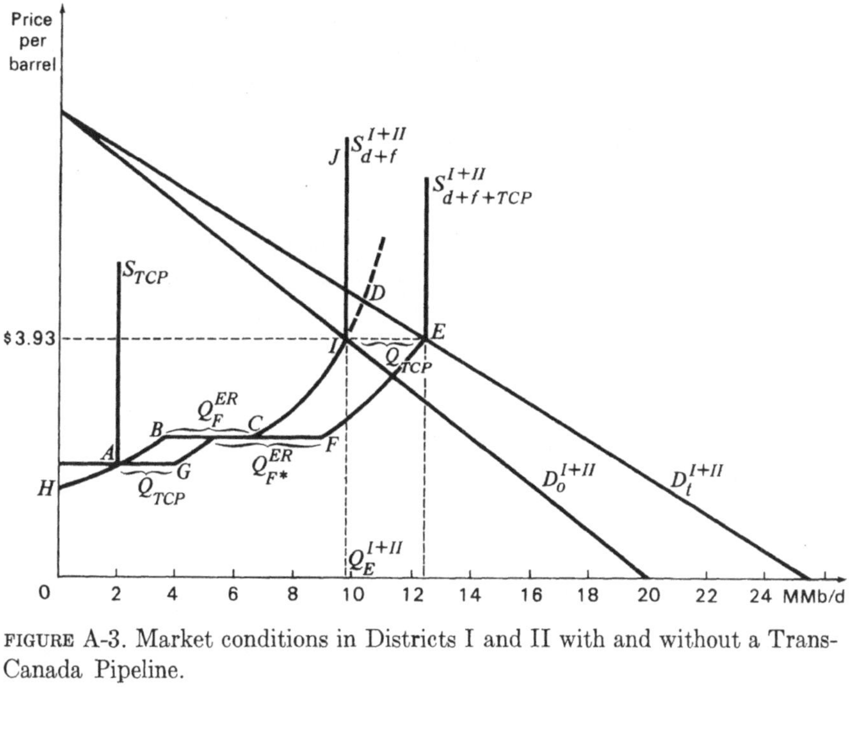

FIGURE A-3. Market conditions in Districts I and II with and without a Trans-Canada Pipeline.

would be derived by horizontally adding the domestic supply curve (with the market demand prorationing restrictions) and the foreign supply curve, which is assumed to include Canadian crude oil and to be based on constant marginal cost. I label this market supply curve S_{d+f}^{I+II}. To avoid confusion, the exact labeling of the various segments that make up S_{d+f}^{I+II} is *HABCIJ*. The fact that domestic supply is restricted to the equilibrium quantity, Q_E^{I+II}, minus the quantity of foreign oil, Q_F^{ER}, results in an inelastic supply curve for both domestic and foreign oil at the equilibrium level, as indicated by the *IJ* segment.

If assumptions similar to the TAP examples are made for a direct North Slope to Midwest Trans-Canadian Pipeline (TCP), their meaning can be shown in figure A-3. First, assume that the supply curve for TCP is represented by S_{TCP}, that is, there are constant marginal costs, but the quantity of oil supplied is restricted by the available throughput capacity to a quantity equal to Q_{TCP}. Second, assume that the throughputs of TAP and TCP are equal. Third, for comparability and to avoid increasing the relative benefit estimates of TCP, assume that the quantity of foreign oil (Q_F^{ER}) before the North Slope oil is added to supply does not increase.[3] Fourth, the demand curve in Districts I and II, the Midwest and East Coast, shifts from D_o^{I+II} in the present to D_t^{I+II}. For simplicity assume that the quantity of oil demanded increases by an amount equal to Q_{TCP} at the initial equilibrium price of $3.93. This price is based on an average of the 1971 price in the Midwest and on the East Coast for crude similar in specific gravity to North Slope crude, namely 26°–26.9° API.

By summing the TCP supply curve, S_{TCP}, and the Districts I and II restricted domestic and foreign supply curve, S_{d+f}^{I+II}, horizontally, the supply curve in the Midwest and East Coast markets can be derived using assumptions one to three above. Let segment *ID* be the alternative additional supply that would be produced from either domestic or foreign sources if TCP were not built. The benefits (before subtracting any external effects) of a TCP could be estimated by the irregular area labeled *ABCDIDEFG*. In chapter 3 I use a benefit estimation methodology that is based on the producers' surplus portion of these benefits; that is, an area similar to *IDE* is excluded from that analysis. It is considered, however, in chapter 4.

[3] One would actually expect this quantity to increase, since domestic production increases (using the present import quota formula) would result in a 12.2 percent increase in the level of imports. Such an effect would increase the benefits of TCP relative to TAP, since it would substitute lower-cost foreign oil for domestic oil.

Market Composition

In tables A-1 and A-2 I show the actual effect of such market restrictions on the quantity of oil supplied to these different U.S. markets in 1970 from different sources. The somewhat surprising result is that, with very different formulas for setting the level of imports, the ratio of imports to domestic production[4] was the same in both regions of the country in 1970 at 12.2 percent.

Nationally the actual imports of oil were slightly more than 30 percent of domestic production in 1970. Only a small percentage (4.2 percent) of domestic demand in 1970, however, came from North Africa or the Persian Gulf. These two areas are considered to be politically insecure—a statement that is not made to minimize a possible increasing dependence on these sources of petroleum. In fact, the present known reserves of liquid petroleum in the Western world are dominated by these two sources of supply. Accordingly, as the demand for petroleum products grows in the United States (table A-3 shows some recent

[4] Including the Trans-Mountain throughput, which is considered safe for national security purposes.

TABLE A-1. District V Sources of Supply in 1970

(thousands of barrels per day)

	Daily total	Cumulative daily total
District V production	1,304	1,304
Other domestic sources	193	1,497
Canadian overland exemption	220	1,717
Venezuela	100	1,817
Persian Gulf	115	1,932
Indonesia	100	2,032
Less exports to Southeast Asia (Taiwan)	−78	1,954
Ratio of net foreign imports less 50 barrels of Indonesian supply to District V plus Canadian exempt supply	$\frac{(100 + 115 + 50 - 78)}{(1{,}304 + 220)} = 12.2\%$	
Total actual imports (exc. Canada)/(Domestic production in District V + Canadian production)	$= \frac{237}{1{,}524} = 15.6\%$	
Total actual imports/(Domestic production in District V)	$= \frac{457}{1{,}304} = 35.0\%$	

SOURCE: Bureau of Mines, Office of Oil and Gas, *Summary of International and Interregional Flows of Crude Oil in 1970* (1971).

TABLE A-2. Sources of Supply in Districts I to IV, 1970

(thousands of barrels daily)

	Daily total	Cumulative daily total
Districts I to IV production	10,007	10,007
Other domestic sources	24	10,031
Outflows to District V	−193	9,838
Canada (includes 35 Mb/d above the 1970 annual limit set by President Nixon)	430	10,268
Mexico	45	10,313
Caribbean (includes 34.8 Mb/d to Puerto Rico outside the system)	1,550	11,863
Other Western Hemisphere	225	12,088
Free Europe	120	12,208
North Africa	220	12,428
West Africa	80	12,508
Middle East via pipeline and tanker	25	12,533
Middle East via the Cape of Good Hope	260	12,793
Sum of foreign imports less 1,513 Mb/d in residual fuel oil, less Canadian above the limit (35 Mb/d), less exports (142 Mb/d) and less Puerto Rico (34.8 Mb/d) divided by domestic production in Districts I to IV	$\frac{1{,}222}{10{,}007} = 12.2\%$	
Total actual imports (Domestic production in Districts I to IV)	$\frac{2{,}955}{10{,}007} = 29.5\%$	

SOURCE: Bureau of Mines, Office of Oil and Gas, *Summary of International and Interregional Flows of Crude Oil in 1970* (1971).

trends), the prospect for turning to such sources becomes more probable over time.

There are two additional reasons for the Persian Gulf area (especially the non-Arab country of Iran) to become a major source of U.S. oil. First, the other industrialized countries of the Western world, most

TABLE A-3. Oil Import Appeals Board Measurement of Domestic Demand

(thousands of barrels per day)

Year	Districts I to IV	District V	National
1968	11,077	1,760	12,837
1969	11,710	1,834	13,544
1970	12,262	1,846	14,108
Rate of change: 1968 to 1970	5%	2.5%	

SOURCE: Bureau of Mines, *Mineral Industry Surveys*, March 18, 1969, March 24, 1970, and December 15, 1970.

TABLE A-4. Production Schedule, California and District V

(*millions of barrels per year*)

Year	California	Total District V
1968	372.3	445.3
1969	375.3	451.9
1970	375.5	457.1
Rate of change: 1968 to 1970	0.5%	1.3%

SOURCE: Bureau of Mines, *Mineral Industry Surveys*, March 18, 1969, March 24, 1970, December 15, 1970, and September 17, 1971.

notably western Europe and Japan, have an even faster rate of growth in demand than the United States, thus relatively secure alternative non-domestic sources of oil cannot be expected to meet this growing demand. Second, production costs in the Middle East are among the lowest, and a growing world oil tanker industry is making advances in

TABLE A-5. 1971 Crude Oil Prices in Various Markets

Market	Quality	Price per barrel (*dollars*)	Source
New York	26°–26.9° API	4.06	*Platt's Crude Oil Summary*, Aug. 25, 1971: South Louisiana (plus gathering costs of 14¢ and transportation of 45¢)
Los Angeles	26°–26.9° API California sulfur range 1% to 2%	3.17	Signal Hill (plus gathering and transportation costs of 5¢)
Chicago	26°–26.9° API	3.81	South Louisiana (plus gathering costs of 14¢ and transportation of 20¢)
South Louisiana	26°–26.9° API (less than 0.5% sulfur)	3.47	(Wellhead price)
Tokyo	Average of all crudes imported	1.83	*Platt's*, Aug. 19, 1971
Canada (eastern)	Average of all crudes imported	1.98	*Platt's*, April 1, 1971
	Venezuela	2.08	

SOURCES: All prices are based on data published in recent editions of *Platt's Oilgram Price Service*, with U.S. prices based on the *Crude Oil Supplement*, vol. 49, no. 164–B (August 25, 1971). Costs for Chicago and New York and the state of Alaska are based on Cabinet Task Force for Oil Import Control, *The Oil Import Question* (February 1970); and for Los Angeles on Arlon R. Tussing, George W. Rogers, and Victor Fischer, *Alaska Pipeline Report: Alaska's Economy, Oil and Gas Industry Development, and the Economic Impact of Building and Operating the Trans-Alaska Pipeline*, ISEGR Report no. 31 (Fairbanks: University of Alaska, Institute of Social, Economic, and Government Research, August 1971).

improved speed, efficiency, and capacity, which tend to reduce the transportation costs of sources of supply in North Africa and the Persian Gulf. Finally, a decision on whether or not to depend on oil from foreign sources is important because host countries to oil companies may increase taxes or demand to participate in, or to take over, ownership and profits of oil operations in the future.

While demand is growing in all districts in the United States, domestic production in California and District V is growing at a much lower rate than in the rest of nation (table A-4 shows rates for California and District V). Therefore if (1) TAP is not built, (2) the Trans-Mountain throughput is not expanded, (3) District V prices remain at their current levels, and (4) there is no change in interdistrict flows of oil in the United States, the most likely alternative source of supply to meet an expected growing demand in District V will be foreign oil, probably coming from the Persian Gulf.

The market prices of crude oil are shown for various sources of supply in table A-5. These relative prices are used in the analysis of different alternative transportation systems in chapter 3. The difference of 89¢ per barrel in crudes of similar quality in Los Angeles and New York should be noted. The different supply restrictions on the West Coast and East Coast are responsible in part for the higher New York price.

APPENDIX B

Summary of Expected Environmental Impacts of the Trans-Alaska Pipeline

TABLE B-1. Possible Effects of the Trans-Alaska Pipeline on Game Resources along the Proposed Route

Species	Population status	Effect
Wolf	Perhaps 5,000 in state, several hundred on North Slope.	Negligible. Some loss of denning sites. Increased hunting likely.
Caribou	Unique. Found only in the Arctic of North America. Over 600,000 in Alaska. Harvests of 30,000/year for food and 35,000/year for sports hunting.	Effects on migration during production from feeder lines depend largely on how lines are designed, constructed, and operated. They should be covered by regulations requiring construction that allows unrestricted movement over or under such structures.
	Three major herds migrate over proposed route. Varying numbers of the largest, or Arctic Caribou, herd of 242,000 winter on North Slope. Porcupine Caribou Herd (140,000) range in the Arctic from Prudhoe to the Mackenzie River. Nelchina Herd (61,000) is south of the Alaska Range.	Compliance with Interior Department stipulation that approximately 50% of the pipeline should be buried should reduce adverse effects to minor ones. Some unresolved questions remain, however.
Moose	125,000 to 160,000 in the state, 2,000 on the North Slope, and an unspecified number along route. 30,000 to 40,000 moose hunters participate each year in Alaska.	Minor effect from pipeline if (1) aboveground pipeline permits unrestricted movement and (2) pipeline trench is backfilled immediately. North of Brooks Range some elimination or reduction of moose is inevitable.
Dall sheep and mountain goat	High sheep populations (several thousand) in portions of Alaska, Chugach Mountain and Brooks ranges. State estimates revenue from some 2,000 sheep hunters in the southern ranges to be several hundreds of thousands of dollars. Aesthetic satisfaction of tourists also important.	Uncertain effect of pipeline on sheep movements and on use of several natural mineral licks. Helicopter noise, and to lesser extent fixed wing aircraft, during preconstruction and monitoring periods may cause animals to abandon their ranges. Hunting will probably increase.

TABLE B-1. continued

Species	Population status	Effect
Wolverine	No population estimates available, but wolverines exist along entire pipeline route. Some 400–700 harvested annually.	Highly unlikely that pipeline would have any measurable effect. Illegal hunting may arise.
Brown grizzly bear		
1) South of the Alaska Range	Occur along entire route, but numbers are unknown. In parts of Europe, Asia, and "lower 48" are considered rare and an endangered species.	If pipeline is constructed so that caribou and moose are not affected, most brown bears, with possible exception of sows with small cubs, should not have difficulty.
2) Between the Alaska and Brooks Ranges	Little information available, but population is believed to be stable.	Garbage may pose a serious problem by attracting bears to camps, where they may have to be killed.
3) North of the Brooks Range	Population believed small relative to land area. Bears here believed to be less productive but to have greater cub survival and to reach an older average age.	Harassment by helicopters and fixed wing aircraft may be harmful during mating season and cause unnecessary death to bears. Oil spills in anadromous fisheries would substantially reduce the bears' high-protein fish diets.
Black bear	Unknown population size, but black bears are common along the route, being more abundant near southern coast and unlikely to be present in treeless northern tundra.	Black bears generally adapt easily to human activity. Since they are omnivores, even major oil spills on land would be unlikely to have a major impact on their population. Garbage from construction and increased hunting may pose problems but probably can be regulated.
Bison		
1) Copper River Herd	About 125 animals.	Adverse effect unlikely.
2) Chitina River Herd	About 16 animals.	Adverse effect unlikely.
3) Big Delta Herd	About 230 animals.	Probable interference from construction crews along the route, and bison are likely to harm themselves by consuming non-edible scrap materials, such as paper, plastic, wood, etc. Crossing the aboveground portions of the pipeline is a problem for this largest herd.
Musk-ox	After a 100-year absence, 48 musk-oxen were reintroduced to their native Arctic slope; 13 more released in 1970. Since it is a unique big game species of North America, there is considerable interest in it.	Impact is likely to come from feeder line and other North Slope construction activities. Aircraft harassment could cause harm. Production activities will probably cause musk-oxen to move to other areas of North Slope.
Raptors (birds of the order of *Falconiformes* or *Strigiformes*)	Heavy use of pesticides has resulted in chlorinated hydrocarbon contamination of raptor prey species and resulted in drastic population declines.	An uncertain effect on nesting areas for some of the 12 raptor species in Alaska along several portions of the proposed route requires additional study.
Peregrine falcon	Population declines have been documented in Alaska and elsewhere.	Present plans call for TAP to run through—and perhaps eliminate—an important peregrine falcon nesting site, where the highway bridge crosses the Tanana River near confluence with Big Delta River.

TABLE B-1. continued

Species	Population status	Effect
Furbearers along the route: Beaver, Mink, Otter, Muskrat, Weasel, Marten, Red fox, Arctic fox, Coyote, Hoary marmot, Arctic ground squirrel, Red squirrel, Lynx	Populations fluctuate with varying predictability.	Promiscuous hunting expected to be a problem. Construction of pipeline and or road near aquatic environment could disturb food chain of beaver, otter, mink, and muskrat. Local denning sites may be destroyed. Major oil spills could have catastrophic direct effect on downstream furbearers such as muskrat, beaver, and mink, or an indirect effect by adversely harming food supply. Pipeline itself not expected to have major adverse effect.

SOURCE: State of Alaska, *Comments on the Proposed Trans-Alaska Pipeline*, State of Alaska Report (July 30, 1971).

TABLE B-2. Effects on Vegetation of Trans-Alaska Pipeline Operation and Construction and North Slope Production

Vegetation resource	Nature of TAP intrusion	Probable effect
Tundra and subarctic vegetation	Pipeline would pass through about 300 miles of tundra and subarctic vegetation that insulates the underlying permafrost.	Once protective mat is removed or seriously disturbed, permafrost thaws rapidly during warmermonths, which leads to impairment of vegetation, drainage patterns, and water quality.
Tundra vegetation is commonly a thick, spongy, matlike growth made up of grasses, sedges, mosses, lichens, and prostrate bushes.	Production on the North Slope (which is mostly a tundra plain), plus feeder lines, gathering facilities, housing, drilling, roads, vehicle movement, etc.	Schofield and Hamilton[a] report Arctic caribou herds depend largely on lichen ground cover during winter months. Lichen destroyed by TAP may take 50–100 years to revegetate.
	Sulfur content of any consequence in North Slope natural gas would cause a certain level of air pollution during production.	Schofield and Hamilton report that sulfur dioxide has an adverse effect on lichens and therefore on the caribou food chain.[a]
All terrestrial and aquatic plants	Major oil spill.	Serious impact on food chain of wildlife that depend on aquatic vegetation in their food chain. Lesser impact on larger migratory wildlife that depend on terrestrial vegetation, since they can avoid areas affected by large oil spills.

SOURCE: U.S. Department of the Interior, *Environmental Impact Statement for the Trans-Alaska Pipeline* (January 1971).

[a] Edmund Schofield and Wayne L. Hamilton, "Probable Damage to Arctic Ecosystems Through Air-Pollution Effects on Lichens," *Proceedings of the Twentieth Alaska Science Conference* (College, Alaska: University of Alaska, July 1970).

TABLE B-3. Intertidal Effects of Terminal and Tanker Traffic on Prince William Sound on Various Wildlife Species

Species and population size	Effect of major spill	Effect of low-level accumulation
Sea otter (2,000–3,500)	Would kill all on direct contact. Since sea otters segregate by sex, population reduction might exceed direct kill.	Uncertain, since sea otters survive in Monterey Bay, California, with low-level exposure.
Stellar sea lion (7,500–10,000)	In highly concentrated area would probably kill all pups, adversely affect all adults present, and drive others from the area.	Probably no direct effect, since sea lions have survived low-level exposure in California. Indirect effect on food supply is possible.
Harbor seal (tens of thousands)	During pupping season, probably would kill pups on contact or by ingestion during feeding. Would kill some adult seals and drive others from the area.	Seals would probably move from the affected area and would be affected indirectly through their food supply.
Other marine mammals: Dall porpoise, harbor porpoise, black fish, and killer whales (no estimate of numbers)	Some would be killed directly by an oil spill; others would be affected through food supply.	Probably no effect.
Marine waterfowl 1) 30 species of *Anatidoe* family of ducks, geese, and swans	A major spill reaching shore during spring and fall migratory periods would be a lethal threat to immense concentrations of waterfowl.	Oil is a known toxic substance for some forms of microorganisms in the food chain of all the waterfowl listed here.
2) Very abundant supply of cranes of the *Gruidae* family (population unknown)	In addition to threat during spring and fall, resident winter populations are especially vulnerable.	
3) Dusky Canada goose (total known world population 15,000–20,000) and trumpeter swan (largest known population 500–700)	A pipeline break near Copper River would seriously affect these populations.	
Deer (not native, annual harvest 500–1,500)	Harm to winter vegetation from large spills would reduce the carrying capacity of the intertidal zone of Prince William Sound for deer.	Deer unlikely to be affected.
Black bear (some unhunted areas have more than 10 bears/sq. mi.)	Would harm vegetation as well as other food sources.	Black bear unlikely to be affected.
Brown bear (healthy and stable population—35 bears are the average harvest)	Direct harmful effect from consuming oil-covered food; indirect effect from destruction of beach vegetation and salmon as food supplies.	Brown bear unlikely to be affected.
Other furbearers 1) River otter and mink	Particularly susceptible since they gather food in estuarine and littoral zones. Destruction of aquatic habitat would be one problem, and because the animals' hair would mat on contact with petroleum, some deaths would be caused by consuming oil during grooming.	Damage related to extent aquatic habitat is reduced. Likely reduction in carrying capacity of area affected.

TABLE B-3. continued

Species and population size	Effect of major spill	Effect of low-level accumulation
2) Scavengers, such as coyotes, foxes, and wolverines	Since these animals occasionally use fringe and beach area for food, they would be harmed by eating dead animals killed by oil spills.	Scavenger furbearers unlikely to be affected.
Miscellaneous avian species 1) Scavengers, such as eagles, sea gulls, crows, ravens, and perhaps hawks	Spills at sea disastrously affect oceanic and littoral birds by soaking them with oil. They die either from ingesting oil preened from their feathers or because oil destroys their insulation. Scavenger birds would probably suffer the same fate if they ate the oil-killed birds.	Effect unknown.
2) Bald eagle	In the United States, the national bird survives in large numbers only in Alaska and would be particularly susceptible to death from scavenging on carcasses of oil-killed marine birds.	Effect unknown.

SOURCE: State of Alaska, *Comments on the Proposed Trans-Alaska Pipeline*, State of Alaska Report (July 30, 1971).

APPENDIX C

A Comparison of the Employment Impact of Alternative Routes on the State of Alaska

by George W. Rogers

The specter of an end to further North Slope development and the capping of existing wells is evoked by proponents of the Prudhoe Bay–Valdez route whenever a Canadian alternative route is suggested. The argument is that any change in plans at this late date would further delay construction, resulting in inflationary cost increases to the point where the companies would be forced to abandon their North Slope prospects. Even if this did not happen, any delay could be damaging to the state. The state's legislative and executive branches have both predicated their spending programs on early construction of the line and rapid development and production from the North Slope fields. With spending already keyed to these high expectations, there is growing fear that at the current rate of living off the capital of the 1969 bonus payments, the state will once more be destitute if production is not forthcoming within the next few years.

This fear is based on an overstatement of the probable future inflationary costs and an understatement of the amount of knowledge already in hand concerning any Canadian alternatives. The cost arguments are furthered by comparing the total mileage of the Valdez and Canadian alternatives. Much of the inflation had been headed off by the pipe supplied in 1969 and early 1970 by Japanese firms. This has all been received in Alaska, coating and other processing has been com-

George Rogers is an economist with the Institute of Social, Economic, and Government Research, University of Alaska.

pleted, and the pipe has been stockpiled at Prudhoe Bay, Fairbanks, and Valdez. Considerable knowledge is already available for designing a Canadian line because of military experience during and after World War II in the construction of highways, supply roads, and petroleum pipelines (from Haines to Fairbanks and from Norman Wells to Whitehorse). Those whose arguments are based on the longer length of the Canadian alternatives generally ignore the added tanker miles and costs of the Trans-Alaska route (TAP) and overlook the long-run economies of a greater throughput on the Canadian line once Canadian Arctic oil is developed.

The most popular arguments for ignoring Canadian alternatives relate to employment. The Alaskan work force is plagued by a chronically high unemployment rate that has increased as a result of the influx of unemployment from the Seattle-Tacoma area. The average Alaskan workingman and voter quickly respond to the argument that any shift of the pipeline to Canada would automatically transfer potential jobs from Alaskans to Canadians. A number of thoughtful analyses have been made recently in Alaska that underline the following points: (1) the employment generated by the pipeline would be only a brief boom (three years with a high peak during the second year and then a rapid drop-off), (2) most of the higher-paying jobs would be filled by imported workers with specialized skills, and (3) the state or those who remain in Alaska after the boom would have to carry the burden of the unemployment and relocation costs when the project was completed. The experience would be a repeat of the defense construction boom of the 1950s and the sharp recession at the end of that decade that bankrupted the unemployment insurance fund.[1] For the longer run, Alaska's economy would be better off if the pipeline construction could somehow be avoided and petroleum production could be developed on schedule. Many Alaskans are still captives of the nineteenth century frontier culture, however, and the short-run prospect of high wages and profits appears more attractive than long-run growth.

A recent comparison of the employment impact of the TAP route (Table C-1) with that of TCP-II (Table C-2) and TCP-IV (Table

[1] See Arlon R. Tussing, George W. Rogers, and Victor Fischer, *Alaska Pipeline Report: Alaska's Economy, Oil and Gas Industry Development, and the Economic Impact of Building and Operating the Trans-Alaska Pipeline,* ISEGR Report no. 31 (Fairbanks, Institute of Social, Economic, and Government Research, 1971); Alaska State Housing Authority, *Community Impacts of the Trans-Alaska Pipeline* (Anchorage, April 1971); and G. Klockenteger, "ALPS Construction and its Impact on the Alaska Economy," Alaska Department of Labor, *Economic Analysis,* vol. 1, no. 2 (April 1971).

TABLE C-1. Total Alaska Employment Attributable to Prudhoe Bay Pipeline Construction (TAP)

(thousands, 12-month average)

Activity	Estimated number employed		
	1973	1974	1975
North of Alaska Range			
Pipeline, road, and station construction	2.8	4.8	4.0
North Slope petroleum production	0.7	1.0	0.8
Other related employment	2.7	4.3	3.2
Subtotal	6.2	10.1	8.0
South of Alaska Range			
Pipeline and station construction	0.5	1.8	1.4
Valdez terminal construction	0.2	0.5	0.4
North Slope petroleum production (Anchorage)	0.7	0.9	0.8
Other related employment	1.6	3.6	3.0
Subtotal	3.0	6.8	5.6
Total Alaska employment attributable to pipeline:			
12-month average	9.2	16.9	13.6
Peak annual employment	13.8	24.0	19.7

SOURCES: George W. Rogers, "Economic Analysis of the Prudhoe Bay-Valdez Pipeline and North Slope Petroleum Development Employment Impacts and Local Hire Implications," submitted to the Joint Pipeline Impact Committees, Alaska Legislature, March 1972. Based on estimates of peak annual employment by Alyeska, July 30, 1971, in Alaska Department of Labor, *Planning Information for Selected Occupations* (November 13, 1971). Peak employment was converted to a twelve-month average by the use of seasonal indexes of construction employment, 1965–71, north and south of the Alaska Range. North Slope petroleum production and other related employment were calculated using the model in Arlon R. Tussing, George W. Rogers, and Victor Fischer, *Alaska Pipeline Report: Alaska's Economy, Oil and Gas Industry Development, the Economic Impact of Building and Operating the Trans-Alaska Pipeline*, ISEGR Report no. 31 (Fairbanks: University of Alaska, Institute of Social, Economic, and Government Research, August 1971).

C-3) showed that a shift to one of these Canadian alternatives would lower Alaskan employment in two of the three years in the TCP-II case but would cause virtually no change in employment in TCP-IV.[2] TCP-II would be routed from Prudhoe Bay across the Brooks Range, then eastward to the border at Old Crow flats—an estimated 285 miles of pipeline with the necessary pumping and control facilities within

[2] George W. Rogers, "Economic Analysis of the Prudhoe Bay–Valdez Pipeline and North Slope Petroleum Development Employment Impacts on Local Hire Implications," submitted to the Joint Pipeline Impact Committees, Alaska Legislature, March 1972.

TABLE C-2. Total Alaska Employment Attributable to Prudhoe Bay Pipeline Construction (TCP-II)

(*thousands, 12-month average*)

Activity	Estimated number employed		
	1973	1974	1975
North of Alaska Range			
Pipeline, road, and station construction	2.8	3.3	2.1
North Slope petroleum production	0.7	1.0	0.8
Other related employment	2.7	2.9	1.5
Subtotal	6.2	7.2	4.4
South of Alaska Range			
Pipeline and station construction	0.7	0.3	—
North Slope petroleum production (Anchorage)	0.7	0.9	0.8
Other related employment	1.6	1.4	0.9
Subtotal	3.0	2.6	1.7
Total Alaska employment attributable to pipeline:			
12-month average	9.1	9.7	6.1
As % of TAP employment	100.0	57.4	44.9
Peak annual employment	12.2	12.8	9.0

SOURCE: George W. Rogers, "Economic Analysis of the Prudhoe Bay-Valdez Pipeline and North Slope Petroleum Development Employment Impacts and Local Hire Implications," submitted to the Joint Pipeline Impact Committees, Alaska Legislature, March 1972.

Alaska. This would eliminate construction of the Valdez terminal and reduce the total line within Alaska by 64 percent. The reduction in employment associated with these changes would be offset in part by the necessity to increase the construction of access roads from 300 miles for TAP to 560 miles for TCP-II and the continued use of the port of Valdez for handling and coating additional pipe required for part of the Canadian route. TCP-II employment within Alaska would be 100 percent of TAP employment for the first year of construction, 52 percent for the second, and 45 percent for the third.

TCP-IV would follow the proposed TAP route as far south as Big Delta and then turn eastward to parallel the Alaska Highway to Edmonton—an estimated 809 miles of pipeline within Alaska, or a 3 percent increase over the TAP mileage. The loss of the line south of the Alaska Range and the Valdez terminal would be offset significantly by the increased pipeline mileage and additional pumping facilities and the extension of the use of Valdez as port of entry for additional

TABLE C-3. Total Alaska Employment Attributable to Prudhoe Bay Pipeline Construction (TCP-IV)

(thousands, 12-month average)

Activity	Estimated number employed		
	1973	1974	1975
North of Alaska Range			
Pipeline, road, and station construction	3.6	7.3	6.2
North Slope petroleum production	0.7	1.0	0.8
Other related employment	3.3	5.9	3.6
Subtotal	7.6	14.2	10.6
South of Alaska Range			
Pipeline and station construction	0.7	0.3	—
North Slope petroleum production (Anchorage)	0.7	0.9	0.8
Other related employment	1.6	1.4	0.9
Subtotal	3.0	2.6	1.7
Total Alaska employment attributable to pipeline:			
12-month average	10.6	16.8	12.3
As % of TAP employment	115.3	99.4	90.4
Peak annual employment	14.7	23.5	19.5

SOURCE: George W. Rogers, "Economic Analysis of the Prudhoe Bay-Valdez Pipeline and North Slope Petroleum Development Employment Impacts and Local Hire Implications," submitted to the Joint Pipeline Impact Committees, Alaska Legislature, March 1972.

pipe and other supplies. TCP-IV employment within Alaska would be 115.3 percent of TAP employment during the first year, 99.4 percent in the second, and 90.4 percent during the third year.

Although the Canadian alternatives would reduce employment within Alaska significantly in one case, they both would result in a concentration of employment north of the Alaska Range where the current unemployment situation is most severe. The TCP-II pattern of employment, which results in a peak level in the first year with progressive declines in the remaining two years, could significantly ease post-construction adjustment costs of the Alaskan economy. A final point is that the most likely level of resident Alaskan participation in these projects would be between 50–75 percent of the total employment generated.[3] This would further minimize the significance of any reduction in Alaskan employment by a shift to a Canadian route.

[3] Tussing, Rogers, and Fischer, *Alaska Pipeline Report*, pp. 80–84.